FIRE PROTECTION

DESIGN CRITERIA, OPTIONS, SELECTION

J. Walter Coon, PE

Illustrated by Carl W. Linde

FIRE PROTECTION

DESIGN CRITERIA, OPTIONS, SELECTION

J. Walter Coon, PE

R.S. MEANS COMPANY, INC.
CONSTRUCTION CONSULTANTS & PUBLISHERS
100 Construction Plaza
P.O. Box 800
Kingston, MA 02364-0800
(617) 585-7880

© 1991

The editors for this book were Mary P. Greene, John J. Moylan, and Julia Willard. Typesetting was supervised by Helen Marcella. The book and jacket were designed by Norman R. Forgit. Illustrations by Carl Linde.

Printed in the United States of America

10 9 8 7 6 5 4 3 2 1

Library of Congress Cataloging in Publication Data

ISBN 0-87629-215-5

Dedication

To my wife and best friend, Patricia, for her patience, encouragement, and thoughtfulness without which I could never have accomplished this task.

To my sons, Anthony and Joseph, and my daughters, Anne and Kathleen, for their continuing expression of faith and pride.

TABLE OF CONTENTS

FOREWORD

There is no mystery to fire protection, only a lack of basic knowledge. Using the information provided in this text, the reader will be able to discuss, on a professional level, a discipline that is a part of almost every construction project.

This book provides a basic foundation in fire suppression, detection, and alarm systems. Fire protection systems, equipment, and devices are discussed, as well as the myriad of jurisdictional individuals, agencies, and standards involved in the field. With this information, architects, engineers, estimators, salespeople, insurance inspectors, safety engineers, firefighters, fire science technology students, specification writers, and university architectural and engineering students will be able to identify particular systems, and understand how they function.

The book was inspired by the author's 43 years working with construction, engineering, and facilities management professionals whose knowledge of fire protection is quite limited. The format presented was adapted from that used in hundreds of seminars given on various facets of fire protection to groups of mechanical engineers, architects, estimators, fire service personnel, designers, and even graduate fire protection engineers who, despite an impressive wealth of education, nevertheless lacked an understanding of the "nuts and bolts" of fire protection system selection, design, and equipment.

As an introduction to fire protection, the student, architect, or engineer will gain an appreciation of the fire protection systems, devices, and equipment available today, and how they can best be applied to a given project.

While fire protection engineering should be left to the experienced professional, those outside this discipline should have sufficient knowledge to not only understand the fire protection engineer's input, but to help make the installation productive and cost effective.

Acknowledgments

The author would like to acknowledge Mary Greene and Julia Willard, the R.S. Means manuscript editors, for their tireless consideration and efforts with my manuscript, spelling, and composition, and for their patience when my manuscript was late.

Appreciation is also expressed to John Moylan, technical editor, for his time-consuming efforts and expertise applied to this manuscript and illustrations, and to the Black and Veatch Engineering Company for offering a retired Fire Protection Engineer the opportunity to continue the practice of his 43 year career in a professional work environment that was the inspiration and catalyst needed to develop this book.

CHAPTER ONE

DEVELOPMENT OF FIRE SUPPRESSION SYSTEMS

CHAPTER ONE

DEVELOPMENT OF FIRE SUPPRESSION SYSTEMS

Early Methods of Suppression

The first recorded attempts at fire suppression involved the use of water as an extinguishing agent to control an uncontrolled fire. Leather buckets filled with water were kept on hand by the householder to extinguish fires in the home, or to wet down combustible roofs to prevent the spread of fire to adjacent buildings from exposure to sparks and radiant heat.

In 1723, Ambrose Godfrey devised an improvement over merely throwing a bucket of water or sand on a fire. Godfrey's arrangement consisted of a barrel filled with water which also held a charge of gunpowder in a waterproof container. The idea was to light a fuse and then throw the barrel into the fire. The fuse ignited the gunpowder and the explosion shattered the barrel, scattering the water on the fire. Crude as it may seem, the Godfrey Barrel was the first attempt to convey extinguishing water to the seat of the fire—an area that previously could not be reached because of intense heat and flames.

Automatic Suppression Systems

The next step in the evolution of fire suppression systems was the automation of water delivery through a piping system, as opposed to applying the extinguishing agent to the fire by hand. The earliest piping system had fittings resembling salt shakers installed in the piping. A cord and weight arrangement controlled the operation of the water supply valve, which was normally kept in the closed position. The water supply was stored in a tank above the piping arrangement, thereby effecting static pressure on the water supply pipe. In the event of a fire, the cord in the fire area burned through, thereby releasing the weights and opening the valve. While the concept was sound, there were several defects. For example, the cords sometimes stretched and lowered the weights, allowing the valve to leak. In other cases, when the system was called upon to function, the valve (lacking modern gasket and lubricating material)—after standing inoperative over a long period of time—would stick and fail to open.

In 1852, the first perforated piping system was installed in America at the Locks and Canals Company in Lowell, Massachusetts. At that time New England was at the center of the textile mill industry. These operations were housed in large, multi-story, wooden structures.

The textile mills contained machinery driven by pulley and belting systems, plus primitive heating and lighting systems which provided a source of ignition. Once ignition occurred, it could readily feed not only on the wooden construction, but also the lint and fabric products inside. Since these buildings were constructed prior to the age of fire protection engineering and fire-safe construction requirements, when a fire did break out and was not immediately extinguished, it would spread rapidly through unprotected openings to other floors. The loss was usually total. The need for an efficient automatic fire extinguishing system became painfully obvious and economically prudent.

Perforated piping systems appeared to be the most practical answer. The customary installation consisted of manually operated control valves serving risers for each floor. Because of this design, however, the damage was usually severe. For example, a small fire in one corner of a 15,000 or 20,000 square foot floor area meant flooding the entire floor with water. In many cases, the area over the fire might never receive water because perforations would be clogged with paint, sediment, and corrosion.

Development of the Sprinkler

In an attempt to overcome the problems caused by the clogging of holes drilled directly into the perforated piping system, and to afford better water distribution, the next step in the development of automatic fire suppression systems was the installation of crude, open sprinklers connected to the piping. These open sprinklers were metal bulbs with numerous perforations, which resulted in a spray pattern. Unfortunately, these perforations were also conducive to severe clogging, which not only prevented water discharge, but also disrupted the discharge "pattern" the head was supposed to develop.

Fusible Element Sprinkler Head
In London, England, in 1864, Major Harrison developed the first fusible element sprinkler head. Major Harrison's device consisted of a hollow brass sphere that contained a large number of counter-sunk holes plugged with solder. When the solder melted from the heat of the fire, the water was discharged. As with previously described methods, the small holes frequently clogged with sediment, making the head at least partially inoperative. In addition, the Harrison head had a rubber valve which had a very short life, and once inoperative, was useless.

There is no record of a patent in Major Harrison's name for this device, and apparently it was never installed in a system. Perhaps this was because of the defects that appeared in the experimental models. Nevertheless, Harrison's sprinkler represents the first automatic application of water from an individual head. But certainly the reports of this idea influenced future sprinkler head development.

Parmalee Sprinkler Head
It is impossible to say just how much influence the Harrison head concept had on the considerable amount of sprinkler head experimentation that was performed from 1874 to 1878. It was during this period that Henry S. Parmalee invented and developed the forerunner of what we know today as the automatic sprinkler head. The Parmalee Number 3, shown in Figure 1.1, was an upright sprinkler first used in 1875. It consisted of a perforated water distributor completely enclosed in a brass cap which was held in place by solder. When the solder softened from the heat of fire, the cap was forced off by the water pressure, and the water discharged on the fire area.

Parmalee developed this first automatic sprinkler system to protect his New Haven, Connecticut piano factory, and by so doing demonstrated the first practical automatic fire protection system. That same year, Parmalee designed a new model, further improving it in 1878. (See Figure 1.2.) In the following years, many improvements resulted in a mass produced supply.

Skepticism

New inventions are always greeted by suspicion. Parmalee's invention was no exception. Would it work after standing idle for years? Wouldn't water damage be severe if the head operated because of a small fire? If the head or the piping leaked, would the system be capable of controlling a fire of any proportion? Was the expense of installation justified? These questions that were asked in 1878 are still being asked of the systems that are available today. But between 1878 and 1882, the Providence Steam and Gas Pipe Company sold and installed over 200,000 of these crude, but remarkably efficient, sprinkler heads—a testimony to their performance record.

By this time, Factory Mutual Insurance Company (FM) had compiled records clearly indicating the proficiency of automatic sprinkler protection in the facilities they insured. From 1877 to 1888, their records indicated a total fire loss in unsprinklered facilities of $5,707,000, the result of 759

An Early Automatic Sprinkler — The Parmalee No. 3

Figure 1.1

fires, or $7,500 per fire. By comparison, there had been 206 fires in sprinkled properties during their 10-year span, with a total dollar loss of $224,480, or $1,080 per fire. These statistics convinced insurance companies of the dependability of automatic sprinkler protection. The companies, in turn, encouraged their policyholders to install automatic sprinkler systems, thereby protecting themselves against serious fire loss claims. The cost of insurance was reduced if the client installed automatic protection. One example of this sprinkler installation incentive comes from Factory Mutual records which indicate that in 1875, the cost of fire insurance was 30 cents per $100 valuation; in thirty years, this rate had dropped to 4 or 5 cents per $100 valuation.

The Influence of Frederick Grinnell

If one man's name is to be remembered as the father of the automatic sprinkler, it should be Frederick Grinnell. Grinnell was with the Providence Steam and Gas Pipe Company during the Parmalee years. He went on to become President of its successor, the General Fire Extinguisher Company, later named the Grinnell Company—a fire protection company that, to this day, holds a prominent place in the fire protection industry.

Frederick Grinnell was not only the inventor who perfected the Grinnell sprinkler head, but also a man who could see the life safety and property protection value of automatic protection. He used his genius as both an inventor and a businessman to overcome the public's apathy and skepticism, to render automatic sprinkler installations commercially feasible as well as practical.

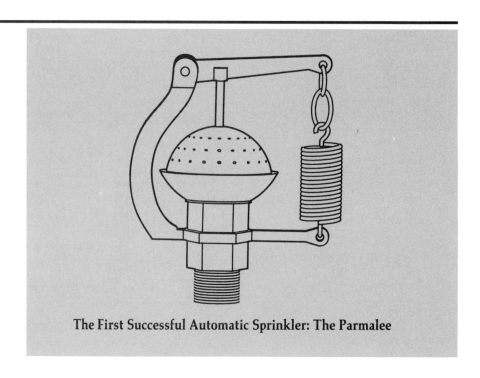

The First Successful Automatic Sprinkler: The Parmalee

Figure 1.2

Between 1873 and 1882 several sprinkler head inventions appeared which influenced the thinking of practical men like Frederick Grinnell. A sprinkler head developed by Charles E. Buell of New Haven, Connecticut in 1873, incorporated a new feature which was to become a standard design feature on all future sprinkler heads. This head used the principle of a baffle, or deflector, to break up the water, effecting both a greater water distribution and smaller water droplets for greater heat absorption.

Kane's Contribution

This era also produced a sprinkler head invented by John Kane in 1881 which used an ordinary ball cartridge as the fusible link. The heat from the fire exploded the cartridge, which in turn released the water. The only problem was that the firing cartridge was positioned with the ball-end pointing downward, so that anyone standing below the sprinkler head became a target.

Between 1872 and 1914, over 450 automatic sprinkler heads were registered with the U.S. patent office, and very likely an equal number were invented but never patented. However, in 1914, Factory Mutual's approval list of sprinkler heads carried only 10 brands, and the 1974 approval list indicated only 15. Today over 15 companies manufacture more than 20 types of sprinkler heads that are approved by nationally recognized testing laboratories such as Underwriters Laboratories (UL) listed and Factory Mutual (FM). Chapter 2 further explains the modern spray sprinkler head and examines some of these devices.

CHAPTER TWO

DEVELOPMENT OF THE SPRAY SPRINKLER

CHAPTER TWO

DEVELOPMENT OF THE SPRAY SPRINKLER

Once practical and efficient sprinkler heads were developed, tested, and approved by a nationally recognized testing laboratory, activity was concentrated on ways to increase the cost effectiveness of sprinkler protection. Research turned toward developing more economical manufacturing methods, less expensive materials, and better quality control. Particular attention was paid to increasing the operating efficiency of sprinkler systems.

Old Style vs. New Spray Sprinkler Head

Sprinkler heads manufactured prior to 1953 are now referred to as "old style" sprinklers because of the drastic industry-wide change that occurred that year with the introduction of the *spray sprinkler*. The old-style sprinkler had an exceptional record of property protection, such that fire insurance companies were cutting the premiums of sprinklered properties as much as 80%. However, construction methods and materials as well as occupancy patterns were changing, and manufacturing and warehousing areas were increasing in size. All of these factors put a tremendous strain on the sprinkler system's ability to extinguish, or even control, high-intensity, large-area fires.

Sprinkler system research at this time concentrated on developing a standard sprinkler head that could control the excessive fire loading created by modern technology and construction, while meeting the increasing demands of industry. The old-style sprinkler head had discharged approximately 50% of the water upward onto the ceiling, which meant that only 50% of the water discharge from the head landed on the fire in a uniform pattern. (See Figure 2.1a.) The objectives of the new research were to discharge more water on the fire without increasing the water pressure and volume, and to break this water discharge up into smaller droplets, thus exposing more water surface to absorb the heat by evaporating to steam.

The new standard spray sprinkler head still used a 1/2" orifice discharge outlet similar to the old-style head. In fact, with few exceptions, all the features of the old style head were similar to the new—except for the deflector, which was completely redesigned to direct approximately 98% of the water downward onto the floor, or fire area. This new type of deflector broke the solid discharge stream through the 1/2" orifice into a water spray consisting of droplets of water smaller than those released by

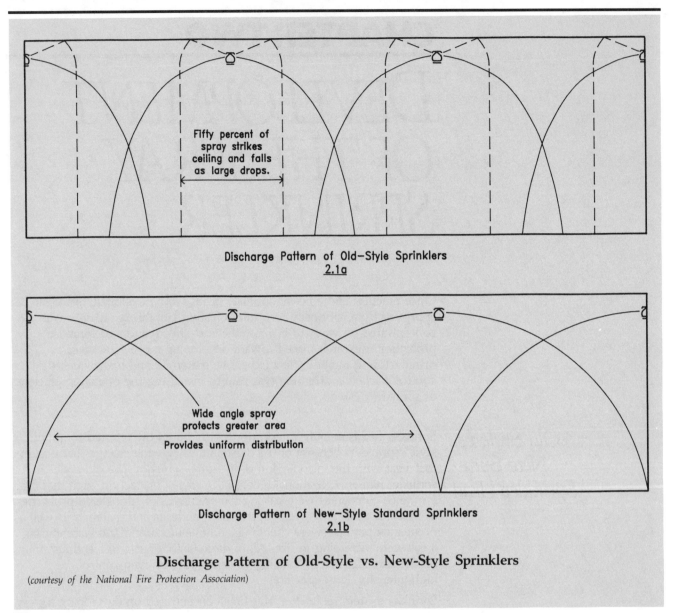

Fifty percent of
spray strikes
ceiling and falls
as large drops.

Discharge Pattern of Old–Style Sprinklers
2.1a

Wide angle spray
protects greater area
Provides uniform distribution

Discharge Pattern of New–Style Standard Sprinklers
2.1b

Discharge Pattern of Old-Style vs. New-Style Sprinklers

(*courtesy of the National Fire Protection Association*)

Figure 2.1

the old style head discharge. Now it was possible, theoretically, to extinguish or control a fire that was producing twice the amount or intensity of heat because the head could now deliver twice the amount of water directly on the fire in a controlled pattern and density. (See Figure 2.1b.)

The introduction of this spray sprinkler allowed an increase in head spacing (square foot coverage per head) because of the expanded and uniform discharge pattern. This meant fewer heads were required to cover the same area. Smaller pipe sizes and valves could be used, which resulted in less total system cost without sacrificing fire protection efficiency. Now that the sprinkler industry had a sprinkler head that discharged 98% of the water onto the floor, a relatively simple mathematical formula could be used to determine the amount of discharge reaching the floor at various head pressures. Pressure versus discharge is discussed in Chapter 8, "Hydraulic Calculations," but it is important to note here that it was the spray sprinkler that introduced to the sprinkler industry the use of hydraulically calculated sprinkler systems.

Sprinkler Head Components

The standard spray sprinkler is "standard" only in its basic design and discharge pattern. As previously mentioned, many companies manufacture sprinkler heads that have been tested and approved by nationally recognized testing laboratories, and every manufacturer has incorporated slightly different features to individualize their own particular sprinkler head. Common to all standard sprinkler heads is the terminology describing its basic components.

The Deflector

The *deflector* is the part of the device against which the water stream is directed. The deflector is usually attached to the frame of the sprinkler head by a deflector screw. This screw not only holds the deflector in place on the frame, but also creates a positive tension on the fusible elements of the head. The pressure is transmitted to the cap, which covers the water discharge orifice, thereby keeping the water from discharging.

The deflector performs two functions: first, it breaks up the solid water discharge after it leaves the opening, or "orifice," into relatively small droplets, and second, it creates the umbrella pattern common to standard spray sprinkler heads. This discharge pattern, and water distribution within this pattern, is relatively uniform at all levels below the sprinkler head. The "old-style" sprinkler head discharge pattern was not uniform at all levels; nor was it uniform within the perimeter of the pattern, due to the discharged water striking the ceiling and falling to the floor. Furthermore, the droplet size was not uniform, as small drops would fall uncontrolled while large drops would fall in the center of the pattern circle.

At a distance of 4' below the spray sprinkler deflector, with the head discharge at 15 gpm (7 psi), the standard sprinkler head covers a circular area of approximately 16' in diameter. With this uniform pattern of water discharge, fire control and extinguishment is greatly enhanced at the floor level, thereby decreasing the exposure of the ceiling to heat from the combustion. In designing the old-style head it had been assumed that it was necessary to wet the ceiling to extinguish the fire. But the research that developed the spray sprinkler proved that more effective extinguishment over a larger area of application could be achieved by directing nearly all of the water toward the floor and into the combustion.

The spray sprinkler head may be used in the upright or the pendent position, but their positions cannot be interchanged. The Spray Sprinkler Upright (SSU) must be used in the upright position, and the Spray Sprinkler Pendent (SSP) must be used in the pendent position, due to the design of the deflector—large on the SSU and small on the SSP. Figure 2.2 illustrates these two types of sprinkler heads.

The Operating Elements

It is in the *operating elements* of the spray sprinkler head that manufacturers express their individuality. The operating elements can be generally classified as *solder-link*, *frangible-bulb*, and *frangible-pellet*.

Solder-link

The solder-link sprinkler consists of a combination of levers, links, and struts arranged so that the water or air pressure acting upward on the cap covering the orifice, and the deflector screw force acting in a downward direction, minimize tension in the solder element. By minimizing the force that acts on the solder, a lesser amount of solder can be used, thereby reducing the level of heat necessary to raise the solder to a temperature at which it becomes soft enough to release the linkage.

The solder used in sprinkler heads is usually an alloy of tin, lead, cadmium, and bismuth, and is prepared to provide a predetermined, sharply defined melting point. Mixing two or more metals that have different individual melting points can produce an *eutectic alloy*, an alloy with a melting point lower than the lowest melting point of any of the metals comprising the alloy.

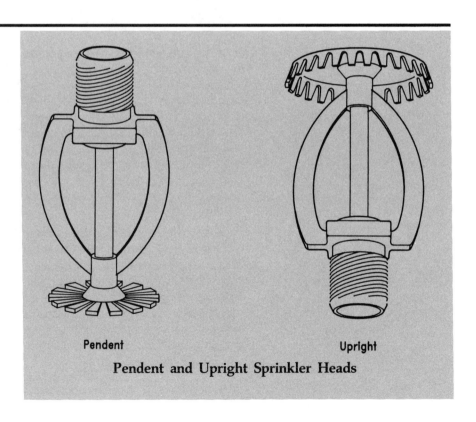

Pendent   Upright

Pendent and Upright Sprinkler Heads

Figure 2.2

The solder, or alloy, in a sprinkler head has a predetermined melting point that establishes the *temperature rating* of the specific head. This temperature rating is stamped on the soldered link. The specific temperature rating of solder-link sprinkler heads is determined by a standard test in which the test head is immersed in a liquid, slowly raising the temperature and noting the temperature at which the head fuses.

Figure 2.3 indicates the temperature rating and corresponding color coding of solder-link sprinkler heads used in areas where the maximum ceiling temperature has been determined. If the range of room temperature is not known, or if a piece of equipment in the area will produce heat above the ambient room temperature, it may be necessary to hang a recording thermometer in the area in question to determine temperature conditions. The recording thermometer is similar to a medical thermometer and will record the highest temperature attained. Unlike ordinary thermometers, it will remain at this temperature until mechanically, or manually, shaken down.

Note that the operating temperature (temperature rating) of the solder-link sprinkler head differs from the maximum ceiling temperature. The reason for this is that solder loses its strength in a temperature range in the "gray area" slightly below its melting temperature. Maintaining a temperature in this gray area could weaken the solder, causing mechanical failure of the head and accidental water discharge. The solder may also be weakened by repeated temperature fluctuations, where the temperature rises to slightly below the melting point and then drops back, and continues to cycle in this manner over a period of time. Therefore, the sprinkler head must have a temperature rating that offers a margin of safety.

The color code indicated in Figure 2.3 is painted on the sprinkler head frame so that an inspector or maintenance worker can identify the temperature rating of the head without having to examine its stamp. The use of color on the head to identify a fusing temperature above the "ordinary" classification is not required for decorative heads, such as those installed below a suspended ceiling, or on chrome-plated heads. Unless otherwise specified or indicated on a plan, the temperature rating of an "ordinary" sprinkler head is 165°F.

Temperature Ratings, Classifications, and Color Codings			
Maximum Ceiling Temperature °F	Temperature Rating °F	Temperature Classification	Color Code
100	135 to 170	Ordinary	Uncolored
150	175 to 225	Intermediate	White
225	250 to 300	High	Blue
300	325 to 375	Extra High	Red
375	400 to 475	Very Extra High	Green
475	500 to 575	Ultra High	Orange

Figure 2.3

The speed with which a solder-link sprinkler head operates depends on many factors. One such factor is *thermal lag*. The heat-absorbing capacity of the metal parts of the head, linkage, and solder can affect the time required for the entire head and linkage to heat sufficiently to raise the temperature of the solder to its fusing point. The larger the amount of solder, the more heat must be absorbed before reaching melting temperature. This is why it is important to minimize the amount of solder required to hold the linkage in tension to a bare minimum.

The temperature differential between the temperature rating of the head and the ambient temperature also affects the speed of operation. A head rated at 212°F, installed in an area where the surrounding atmosphere is 0°F, must absorb considerable heat to overcome thermal lag and reach operating temperature.

Some situations may warrant the use of sprinkler heads with temperature ratings higher than those normally required by the ambient temperatures of the area. Consider the following scenario.

The number of sprinklers that are anticipated to open in a fire is based on several factors: height of the heads above the floor, square footage each head covers and, most importantly, the fire hazard classification of the occupancy. In addition, regardless of the temperature rating of the heads, it is obvious that the heads directly over the combustion will fuse first, and as the heat from the combustion mushrooms at the ceiling, the heads around the perimeter of the combustion will also be activated.

These heads discharge water around the perimeter of the fire (and not directly on the combustion) and are an important part of fire control and property conservation. The water discharge not only "soaks" the combustibles around the fire, but also protects the surrounding material and equipment from damage by the radiant heat of the fire, thereby keeping these items below their ignition temperature.

The heat from the fire may continue to spread across the ceiling, and although the temperature decreases at the edges, it is still sufficient to activate more heads. The discharge from heads that are too far from the origin of the fire to contribute to fire suppression, control, or property protection will only cause water damage and tax the water supply to the operating heads. Therefore, if all the heads in the area are temperature-rated slightly above that required by the ambient temperature of the area, the aforementioned heads on the fringes of the ceiling heat will not fuse as rapidly as heads with a lower temperature rating, preventing unnecessary water damage.

Another danger of heads that fuse too quickly is that they can use up the water supply needed to put out the fire. Chapter 8, "Hydraulic Calculations," describes how extremely important it is to know how many heads will fuse before the water supply is no longer capable of furnishing sufficient water to the system. The chapter on hydraulically calculated sprinkler systems will show the importance of sprinkler temperature rating in preventing heads that are outside of the calculated area from fusing. Establishing the proper temperature rating of sprinkler heads from a chart when the ambient temperature is known, and no unusual heat producing equipment is present, is relatively simple.

It is imperative that the temperature rating of sprinkler heads above the standard requirements be accomplished only by an experienced fire protection professional. Recommending heads with higher temperature ratings than required based on the layman's view of preventing water damage can be very dangerous. These high-temperature heads over the combustion may not fuse while the fire is small and the discharge may come too late to prevent combustion from intensifying and spreading.

High-temperature heads must be used under skylights where the heads are exposed to the direct sunlight or located in unventilated attics, in confined spaces under uninsulated roofs, and in unventilated areas having high-powered electric lights on the ceiling. Specifications usually specifically state that high-temperature heads shall be provided in the area of heat-producing equipment. The specifications also make recommendations as to the location of the head. For example, when protecting a building where a diesel locomotive can enter, sprinklers should not be located any closer than five feet from the center line of the track.

The decision to use higher-than-required temperature-rated heads must be made based on experience and a thorough understanding of fire protection. Otherwise, the time delay in fusing a head, and the opening of too few heads, can allow a fire to develop to a point that is beyond the ability of the sprinkler system to control. A fire that develops rapidly and increases ceiling temperatures quickly will operate a 212°F sprinkler head almost as quickly as a 165°F head, but a fire that develops slowly, with a slow rise in ceiling temperature, will introduce a considerable time differential between the activation of the 165°F and 212°F sprinkler head.

Frangible Bulb Sprinkler Head

The *frangible bulb sprinkler* is similar in construction to the solder-link sprinkler, except for the operating mechanism, which does not depend on the heating and melting of a solder element. The frangible-bulb sprinkler operating mechanism consists of a special glass bulb which takes the place of the mechanical linkage in the solder-type head, but performs the same function of holding the valve cap over the orifice against the water or air pressure. This glass bulb contains a liquid with high expansion qualities. However, the liquid does not completely fill the bulb, allowing for a small air bubble.

When the heat from the fire reaches the head, the liquid expands, compressing the air bubble. If the heat continues to be absorbed by the sealed glass bulb, the liquid continues to expand until it has completely absorbed the air bubble, and the bulb becomes completely filled with liquid. With the cushion of air gone, and the liquid continuing to expand, the pressure inside the bulb increases until it shatters—literally explodes—thereby releasing the valve cap and allowing the water to discharge against the deflector. The air bubble size is a very important design feature of this head, because by adjusting and regulating the ratio of liquid to air when the bulb is sealed, the exact operating temperature of the head can be determined. Figure 2.4 is an example of a typical frangible bulb type sprinkler head.

Frangible Pellet Sprinkler

The *frangible pellet sprinkler* head closely resembles the solder-link sprinkler head in that the valve cap is held over the orifice by a mechanical linkage which is released by the melting of a fusible element of solder or eutectic metal. The primary difference between these two heads lies in the function of the fusible material. The pellet of solder or eutectic metal is in a cylinder under compression from a plunger. When this pellet reaches a predetermined temperature, it melts and is forced out around the plunger due to the compressive force of the plunger. This loss of material (pellet) allows the plunger to move downward and, in so doing, releases the mechanical linkage that has held the valve cap in place. The water pressure then forces the valve cap off, and the water is discharged. Figure 2.5 illustrates sprinkler head components.

Common Characteristics of Spray Sprinklers

As mentioned previously, all manufacturers of sprinkler heads have devised various modifications of the link and lever mechanism, but the devices are all similar in several respects. First, all approved sprinkler heads are constructed so that the mechanical pressure on the valve cap far exceeds any anticipated fluid pressure exerted on the bottom of the cap. This feature is incorporated into all sprinkler heads to prevent any possibility of a severe pressure surge, water hammer, or exceptionally high water pressure and the leakage around the valve cap that may result.

Another feature common to all sprinkler heads is the stamp required on the metal piece of the fusible element stating the temperature rating of the head. In addition to this, every head will have the use-position stamped into the deflector, frame, or fusible link. The manufacturer's name and model number are cast into the frame, and all approved heads will indicate the date of manufacture stamped into the fusible element.

Frangible Bulb Sprinkler Head

(courtesy of Grinnell Fire Protection Systems Company Inc.)

Figure 2.4

The frame of most standard sprinkler heads is made of high-grade bronze, and the valve cap or disc covering the water entrance orifice is usually constructed of copper or bronze, or a combination of these two metals. One manufacturer features a copper disc in the cap which forms an expansion chamber between itself and the bronze cap. In this way, if the water should freeze, thereby causing expansion pressure, the copper disc would be punctured by the pressure, and a warning drip of water rather than a complete water discharge would occur to indicate the frozen condition.

Development of the Large Orifice Head

The new spray sprinkler was a solution to the threat of the "high-challenge" fire posed by industrial advancement. It was also an answer to better fire protection at a smaller investment. Nevertheless, the interest in improved fire protection was not dulled by this radical change in sprinkler head design. High-level fire challenges—fires that were beyond the capabilities and discharge capacities of standard spray sprinkler installations—were continuing to develop at an even pace with industrial advancements.

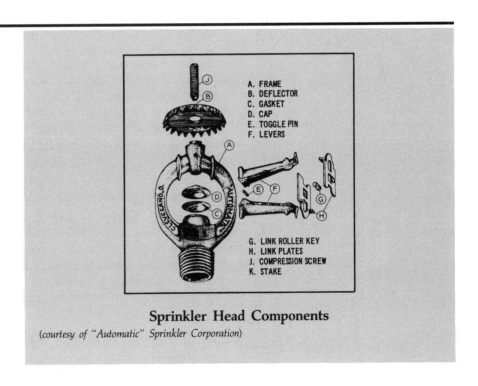

Sprinkler Head Components
(courtesy of "Automatic" Sprinkler Corporation)

A. FRAME
B. DEFLECTOR
C. GASKET
D. CAP
E. TOGGLE PIN
F. LEVERS

G. LINK ROLLER KEY
H. LINK PLATES
J. COMPRESSION SCREW
K. STAKE

Figure 2.5

When an intense fire develops, it can create exceptionally strong updrafts, or fire plumes. As a result, the discharge from standard sprinklers operating over such a fire may be unable to penetrate the fire plume, with the droplets turning to steam far above the burning fuel surface, or combustion core. With the expanded spacing that the spray sprinkler allows, an excessive number of sprinklers around the perimeter of the fire center will begin to open when the first sprinklers directly over the fire fail to control the combustion. As more and more heads open, the discharge density of the heads directly over the fire will decrease, producing less efficient fire control.

Modern technology demanded a sprinkler head with a higher water discharge and density. It also required sprinklers with droplets in the discharge large enough to penetrate the kinds of fire plumes found in intense fires from combustible products that were not anticipated when the spray head was first installed. The improved sprinkler head would also have to cope with fire plumes resulting from storage configurations of a height and extensive square footage area unheard of 10 or 15 years ago. Development of the *large orifice head* was the result: a sprinkler head similar in all respects to the 1/2" orifice sprinkler with one design exception—a 17/32" orifice.

It was discovered that an orifice 1/32" larger than the normal 1/2" orifice would deliver 140% more water, *at the same pressure*. This proved exceedingly advantageous when there was a fire load demand for high water density without the corresponding demand for high water pressure to meet the density requirements. The fire protection engineer could now meet the high density gpm per square foot requirements of rack storage, plastic and rubber storage, flammable liquid operations, and a multitude of situations occasioned by new products and processes, and changing industrial storage and operational demands. These needs could be met with smaller pipe sizes and less pressure than the 1/2" orifice head required. Chapter 8, "Hydraulic Calculations," describes in detail the importance of demand and supply conditions, and points out the advantages of the large orifice head over the regular 1/2" size in satisfying unusual demand and supply situations.

The pipe thread connection of a large orifice sprinkler is 3/4" NPT (National Pipe Thread). The standard 1/2" orifice sprinkler has a 1/2" NPT pipe thread connection. This explains why the large orifice head is sometimes referred to as the 3/4" head.

Grinnell Fire Protection Systems Company makes an interesting adaptation of the large orifice sprinkler head. The orifice is the 17/32" standard, but the pipe thread connection is only 1/2" NPT. With the proper hydraulic calculations and complete re-evaluation of the system layout, it is possible to remove the standard spray sprinkler (1/2" orifice) from an existing system and to install this Grinnell head in its place.

Small Orifice Head

While the large orifice head is used where a high density (gpm per square foot) is required and/or the residual pressure (the pressure "left over" which shows on the gauge, when water is flowing) is low, the small orifice head is used when low density is required and/or a high residual pressure is available. Smaller orifice heads discharge less water than the standard 1/2" orifice head at the same pressure. They are used in both light and ordinary hazard areas where there is a limited water supply, for conservation and efficient limitation of water. The application of the small orifice head is limited and must be recommended with great care by experienced fire protection engineers, and only after all of the protection circumstances have been carefully evaluated.

The small orifice head is available in several orifice sizes: 1/4", 5/16", 3/8", and 7/16". When considering the use of the small orifice head, one should keep in mind that increasing the 1/2" orifice head by 1/32" increases the discharge by 140%. For an example, at 65 psi at the sprinkler head, the 1/2" orifice head will discharge about 45 gpm, the 17/32" orifice head will discharge about 66 gpm, and the small orifice 1/4" head will discharge about 13 gpm.

To avoid mistaking a small orifice head for a standard 1/2" head, every small orifice head with an orifice less than 1/2" has a very noticeable *pintle,* or extension of the compression screw, which extends 1/2" to 3/4" above the deflector. In this way, an inspector can tell at a glance which size sprinkler head he or she is dealing with.

Dry Pendent Sprinklers

A dry pipe system is installed in areas subject to freezing. Water is only admitted to the system piping when a head fuses and the compressed air that holds the water out of the piping is exhausted through the open head.

Once a dry system has "tripped," or filled with water, the open head or heads must be replaced, and the system must be drained of all water and refilled with compressed air to prevent the water in the pipes from freezing. A review of a cross-section of a sprinkler head will reveal a pocket from the valve cap on the orifice to the end of the male pipe thread where the head screws into the pipe fitting, and where water could collect if the head was installed in the pendent position. This water would remain when the dry system piping was drained, and would be subject to freezing.

Sprinklers fed from wet pipe sprinkler lines above the suspended ceiling consist of standard pendent sprinkler heads installed on 1" pipe drop nipples. The pipe drop nipples are cut to fit at the job site so that the pendent head can be installed tight to the suspended ceiling. The concealed wet pipe system is installed before the suspended ceiling is put in place, and the drop nipples that will eventually supply the heads below the ceiling are installed in the line tee outlets. These are "rough-cut" drop nipples, extending 6" or 8" below the future ceiling elevation. Rough-cut drop nipples are usually plugged so that the piping can be hydrostatically tested for leaks before the ceiling is in place. A mark is made on the rough drop nipples at the point where they penetrate the ceiling, and they are then "backed-out" of the tee. They are cut off at the proper length, threaded, reinstalled in the line tee outlet—complete with a 1" × 1/2" reducing coupling on the ceiling end. The pendent sprinkler is then screwed into the 1/2" outlet.

When the sprinkler system piping is installed above a suspended ceiling (concealed piping), and the sprinkler heads are all that show below the ceiling (down feed), the drop nipple from the sprinkler line to supply these pendant heads below the ceiling will obviously fill with water when a dry system trips, and will not drain when the dry pipe system is emptied of water. Industry solved this problem with the development of the *dry pendent sprinkler* (See Figure 2.6).

A dry pendent sprinkler head consists of a drop nipple with a standard pendent head sprinkler attached. Inside the drop nipple is a specially designed shaft that holds a cap at the top of the nipple to prevent water from entering the drop nipple. This shaft is under tension in order to hold the cap against the water pressure. This tension is maintained by engaging the shaft with the operating elements of the sprinkler head. When the sprinkler head fuses and the operating elements are released, the restraint on the shafts is relieved. It then drops a predetermined distance, which pulls the cap away from the opening, allowing water to enter the nipple, flow around the sides of the shaft, and discharge through the orifice of the head.

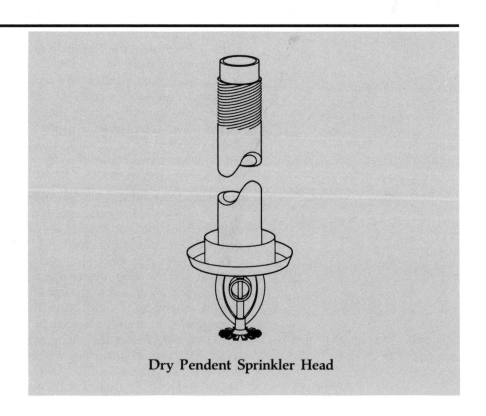

Dry Pendent Sprinkler Head

Figure 2.6

Unlike the standard pendent sprinkler heads, the dry pendent sprinkler cannot be "cut-to-fit" in the field due to the single unit construction of the sprinkler head and drop nipple. After the concealed line piping of a dry pipe system is installed, accurate measurements must be taken from the face of the fitting (the bead of the line tee 1" outlet) to the underside of the suspended ceiling. These dimensions are then used to order the dry pendent sprinkler heads.

Note that when the dry pendent head is screwed into place, it projects a considerable distance into the fitting. For this reason, the dry pendent sprinkler must always be connected to a tee and never to an elbow. If used in an elbow, the sprinkler head could impinge on the curvature of the fitting, and could also create turbulence that could effect the flow of water from the fitting to the sprinkler head. Therefore, when dry pendent heads are installed on the end of a sprinkler line, the elbow must be replaced with a tee and nipple and cap, with the dry pendent sprinkler installed in the tee outlet.

Dry pendent heads can also be used on wet pipe systems to protect a freezer or cooler. For example, a supermarket may be protected with a wet pipe sprinkler system because the store is heated and not subject to freezing. The insurance carrier, however, might require protection of a small freezer box where a pendent sprinkler supplied from the wet pipe system would surely freeze. A dry pipe sprinkler in the freezer box supplied by the wet pipe sprinkler piping would be appropriate in this case.

In this example, if the top of the freezer box were several feet below the ceiling where the wet pipe system piping was installed, it would be necessary to drop down from the ceiling with the supply pipe for the dry pendent head, and install the horizontal tee and nipple and cap for the dry pendent head a few inches above the top of the freezer box. Dry pendent heads are expensive, and the cost of the head is based on the length of the dry pendent head unit. The same design criterion is true for dry pendent heads on a dry pipe system. The pipe must be installed as close to the suspended ceiling as possible, because once a dry pendent head unit exceeds the 6" base cost, every inch in length adds to the cost.

Sidewall Sprinklers

Sidewall sprinklers are similar to standard sprinkler heads, with the exception of the deflector. The deflector produces a discharge that looks like half of a normal sprinkler umbrella pattern. Installed on one wall of a room, a sidewall head will distribute, or "throw," this pattern across the length of the room, covering the entire square footage of the room. A typical sidewall sprinkler is shown in Figure 2.7.

Naturally there are limits to the discharge distance of regular sidewall sprinklers, and the NFPA in their sprinkler standard mandates that if the room or bay exceeds 15', sidewall sprinklers shall be installed on both sides or walls of the area. If the room or bay exceeds 30', NFPA mandates that additional sprinklers with regular patterns shall be installed.

Sidewall sprinklers have many applications, but only in what are referred to as *light hazard occupancies*. As the name implies, these are hazards that do not present the probability of intense fires because the quantity of flammable combustibles is "light." Light hazard occupancies include hotels, nursing homes, offices, and so forth. Light hazard occupancies do not include the storage areas or mechanical rooms that are part of the hotels, nursing homes, and offices. These are classified one step higher as *ordinary hazard areas.*

There are currently on the market sidewall heads called *extended coverage heads* that are designed for light hazard areas where extra-long water spray patterns are needed; for example, in above-average sized rooms in hotels, nursing homes, motels, and so forth where standard sidewall heads are impractical. The extended discharge sidewall head patterns are mentioned here for reference only, because their installation and use must be overseen by experienced fire protection engineers and sprinkler designers.

Sidewall sprinklers are available in the upright, pendent position, as well as in the horizontal positions. They are installed along sidewalls, where their deflector discharges the majority of the water away from the wall, with only a small portion discharged against the wall behind the sidewall head.

Upright and pendent sidewall sprinkler deflectors should be positioned at a distance from walls and ceilings not more than 6″ or less than 4″. Horizontal sidewall sprinklers are allowed to be positioned less than 4″ from the wall, and 6″ to 12″ below noncombustible ceilings.

The following example demonstrates the advantages of sidewall sprinklers. Installing a sprinkler system in an existing hotel where the ceilings are plastered, and installing piping above the ceilings with pendent heads would require tearing out the plastered ceilings and replacing them. By using sidewall heads and installing the sprinkler piping close to the ceiling and close to one wall of the rooms where it will hardly be noticeable, the rooms can be protected without disturbing the existing ceilings.

Sidewall, Horizontal Sprinkler Head

Figure 2.7

This is but one of hundreds of examples in new and existing construction where sidewall heads can provide adequate protection. However, these special discharge heads must only be used under the recommendation of an experienced engineer, so as not to stretch their protection parameters beyond their limitations.

Decorative Sprinklers

The sprinkler industry has responded to the restrictions by architects that regular pendent sprinklers showing below their suspended ceilings are too unsightly. The result—the industry has provided a vast array of decorative sprinklers.

In designing a more attractive device, the sprinkler industry knew that as long as the deflector of the pendent head was one inch below the ceiling when the head discharged, they could continue to provide a suitable discharge pattern. With this in mind, they proceeded to conceal as much of the body of the sprinkler head as possible.

In preparing a specification for the use of a decorative sprinkler head, the writer must be careful not to use generic terms like *flush head*. It is imperative to identify the type of head being specified by a manufacturer's name and model number—or engineer approved equal. The reason for this is best explained by browsing through the illustrations of decorative heads. Some manufacturers refer to the *recessed head*—which is just a pendent head recessed in a cup-type ceiling escutcheon—as a flush head. Figures 2.8 and 2.9 illustrate these two very different types of decorative heads.

There are decorative heads available that are completely concealed, where the only part that shows below the ceiling is a round plate. The plate is held in place by a fusible material that will melt before the fusible element of the sprinkler head melts. When this happens, the plate will drop off exposing the sprinkler deflector. The deflector, in turn, drops an inch below the ceiling when the head fuses.

Naturally, decorative heads are more expensive than regular brass sprinkler heads, and the more complicated or the more concealed they are, the more expensive they will be. The use of chrome pendent sprinkler heads, which are slightly more expensive than brass pendent heads, may prove an acceptable alternative to decorative heads. So will the recessed pendent head. However, in areas where appearance is extremely important, the additional cost of a truly flush, or totally concealed head may well be worth the additional cost.

Special Heads

There are a number of special heads available today. They are briefly described here for reference purposes only, as they have very limited applications and should only be specified and installed by experienced fire protection designers.

Quick Response Sprinkler

A sprinkler head with a temperature rating of 165°F will operate when the head reaches a temperature of 165°F, plus or minus 5°F. However, because of the heat absorption of the mass of the fusible link, the sprinkler head frame, the piping, etc., the air temperature may be as high as 1,000°F before the fusible element operates. This phenomenon is referred to as *thermal lag*.

Recessed Sprinkler

(courtesy of Grinnell Fire Protection Systems Company Inc.)

Figure 2.8

Flush Sprinkler

(courtesy of the Viking Corporation)

Figure 2.9

In response, the industry has developed a sprinkler head with an extremely sensitive fusible element that overcomes most of the thermal lag. As a result, a 165°F head will operate approximately six times as fast as a sprinkler head with a conventional fusible element temperature rated at 165°F. These heads are called *quick response sprinklers*, and have many uses, but the layman should leave their application to experienced fire protection professionals.

Large Drop Sprinkler

The state-of-the-art system for storing commodities is the use of rack storage, with some system racks reaching heights of 100'. A fire in a rack storage warehouse produces extremely high heat release and rapid fire development that can easily overpower the heavy discharge of a sprinkler system. The narrow aisles and flue spaces between the racks create a "chimney effect," with heat plumes racing up the face of the rack storage at speeds exceeding 30 miles per hour.

Spray sprinklers installed at the ceiling and in the racks themselves produce large volumes of very fine droplets of water. However, tests have proven that the majority of these droplets—up to 98%—are unable to penetrate the high velocity heat plumes. In response, the industry has developed a new head called the *large drop sprinkler* which, as its name implies, produces very large drops in the discharge. Tests with these large drop sprinklers indicated that over 140% of the water discharged penetrated the heat plumes and extinguished the combustion. The balance of the discharge from these large drop heads performed cooling that assured control of the combustion until the fire was suppressed.

Early Suppression Fast Response (ESFR) Sprinkler

Factory Mutual Research combined the advantages of the quick response sprinkler head with the advantages of the large drop sprinkler, and the result was the *early suppression fast response*. ESFR heads provide the protection that their name implies: They respond much more quickly than regular heads with the same temperature rating, and discharge a density with drop sizes that provide fast suppression of combustion.

These ESFR heads offer tremendous advantages in suppressing the high-challenge fire, especially rack storage fires, but systems using these heads require a high level of fire protection engineering expertise and should never be specified or recommended by the layman.

On-Off Sprinkler

Another special sprinkler head is the *on-off sprinkler*. This head automatically resets itself after a fire has been extinguished. Water is discharged when the temperature rating of the head is reached; when heat from the fire drops below a predetermined temperature, the head automatically turns off.

One manufacturer describes the operation of its on-off sprinkler as such: When a bi-metallic snap disc is exposed to heat in excess of 165°F, it snaps open and allows a small amount of water to pass through a small pilot orifice. This passage of water releases pressure to a piston assembly, forcing the piston down and allowing the water to be discharged through the main orifice or port. When the fire is controlled and the heat decreases, water pressure builds up in the pressure chamber and forces the piston assembly closed. The discharge flow of water is arrested, and the sprinkler head is therefore automatically reset when the snap disc returns to the operational position.

These on-off heads should be used with great care and only with the approval of the building's insurance carrier. When a standard sprinkler head fuses it remains open until someone in authority determines that the fire has been extinguished and the water supply to the system may be closed and the fused head replaced.

Insurance carriers want the reliability of a standard sprinkler system, as well as a physical assurance that the fire is out. They will permit the use of these on-off heads only for a very special condition, and even then they will limit the number of on-off heads used to a particular area or hazard.

Corrosion-Resistant Sprinklers

Corrosion-resistant sprinklers also fall under the special head category. These are standard sprinkler heads that have been coated at the factory with wax or lead to prevent the head from corroding. When a sprinkler system is installed in an area containing chemicals, moisture, corrosive vapors, or where the heads are exposed to the elements, corrosion-resistant heads must be installed.

According to the specifications, the corrosion-resistant coatings shall be applied only by the manufacturer of the sprinkler head. Any homemade coating application will very likely render it an unapproved head. Furthermore, if the protective coating is damaged during installation, it must be repaired using only the coating furnished by the sprinkler manufacturer and applied in an approved manner, so that no part of the head is exposed to the corrosive atmosphere.

Rack Sprinklers

The sprinkler heads used in storage racks also fall under the special head category. These are standard heads with one exception; they have a water shield about three inches in diameter attached above the deflector on upright heads, and attached above the frame of the head on pendent heads. The reason for this is simple.

These sprinklers are installed in storage racks at various levels of the rack—this is in addition to a sprinkler system at the ceiling of the rack storage warehouse. When a sprinkler at the ceiling discharges, or when a head located higher up in the rack discharges, it will throw water on the heads at lower elevations. This water would keep the heads below the discharge wet and cool so that they cannot absorb sufficient heat to fuse. This phenomenon is called *cold soldering*. The water shield on these lower heads keeps the water discharge (from above) from wetting and cooling the fusible links of the lower heads. In short, they act as "umbrellas" for the heads to keep the fusible links dry.

It is important to note here that cold soldering does not occur in rack sprinkler systems alone. Sprinkler heads on any sprinkler system must be installed no closer to one another than 6' to prevent cold soldering. *On standard, horizontal, sprinkler systems, never install sprinkler heads closer together than 6', or one head discharge will prevent another head from fusing.*

Since a rack sprinkler with a fixed water shield is more expensive than a regular head without the water shield, there are on the market separate water shields that can be attached in the field to regular upright and pendent sprinkler heads. These separate water shields come with a *sprinkler head guard*, which is simply a wire enclosure around the head. In rack sprinkler systems, where commodities are constantly being added or removed from the racks, the probability of damaging the head, especially the fusible link, is very high. The head guard protects the rack sprinkler head from physical damage.

Head guards should be used on any sprinkler system where the heads are subject to physical damage or are within reach. Head guards can also be obtained for some decorative heads.[1]

Sprinkler Accessories

Dust Caps

If sprinklers are installed in areas where severe dust conditions exist, sprinkler dust caps can be used to cover pendent heads. These dust caps prevent dust from accumulating on the head and slow down the fusing of the fusible element.

The dust cap is held in place with solder. The solder is formulated to melt at a temperature well below that of the temperature rating of the sprinkler head, whereupon it will fall away and expose the head.

Spare Heads

Every sprinkler installation must have a supply of spare sprinkler heads in every type and temperature rating used in the system. The spare heads are to be installed as quickly as possible so that the system may resume service without delay. The NFPA dictates how many spare heads shall be on hand based on the number of heads in the system. These spare heads are to be stored in a cabinet designed specifically for that purpose. Contained within the cabinet will also be a sprinkler wrench which is designed to fit the heads installed in the particular system. This cabinet is usually installed adjacent to the sprinkler riser.

With an understanding of the new vs. old type sprinkler head, and a basic idea of the sprinkler heads available today, we now turn to Chapter 3, an explanation of the sprinkler systems in which these heads are used.

[1] *See the Appendix for a chart showing sprinkler head types, characteristics, usage K-factors, NFPA reference, and restrictions.*

CHAPTER THREE

WATER AGENT SUPPRESSION SYSTEMS

CHAPTER THREE

WATER AGENT SUPPRESSION SYSTEMS

There is nothing new in the world, it has often been said: The same basic elements and concepts are merely used in different ways. This holds true for the wet pipe sprinkler system in the use of water as the agent in fire protection. All other automatic fire protection systems using water as the suppression agent have evolved from the wet pipe system. The wet pipe system has been sophisticated and adapted to specific requirements and environments, therefore, a thorough understanding of this system is the foundation on which to build basic knowledge of other systems using water as the fire suppression agent. After explaining the wet pipe sytem, this chapter will describe other systems in common use today: dry pipe sprinkler systems, pre-action systems, deluge systems, standpipe systems, water spray systems, and foam systems.

Wet Pipe Sprinkler System

The definition of a wet pipe sprinkler system in the National Fire Protection Standard 13, titled "Installation of Sprinkler Systems" reads: "A system employing automatic sprinklers attached to a piping system containing water and connected to a water supply so that water discharges immediately from sprinklers opened by fire".

The wet pipe sprinkler system above ground is the consideration in this section, since the water supply and underground part of the system will be covered in other sections.

The purpose of a sprinkler system is twofold: detection and extinguishment. Detection is accomplished when a sprinkler head fuses and water flows in the system, which activates a water flow alarm device which, in turn, activates the alarm system. The discharge from the fused sprinkler head provides the fire suppression. (See Figure 3.1.)

Extinguishment is obviously with water. Wet pipe sprinkler system piping is constantly and completely filled with water under pressure, and when the sprinkler head is heated by a fire condition to the pre-set temperature, the fusible element releases the cap which encloses the water orifice, and water is *immediately* discharged onto the fire. This is one very important advantage of a wet pipe system—*there is no delay* in automatically applying extinguishing water after the sprinkler head has determined that the fire is of a magnitude requiring immediate suppression.

The piping in a wet pipe sprinkler system is sized to supply water, at the required volume and pressure, to a pre-determined number of *heads*. The number of heads anticipated to operate in a fire condition is based on studies and calculations of the fire spread in various occupancies and types of buildings. Chapter 8, "Hydraulic Calculations," shows how the pressure at each sprinkler head is affected as more and more sprinkler heads become operable and their combined flow demands more and more water from the supply source through the supply piping to the operating heads.

If the fire should gain unanticipated headway, and the fire area extends beyond the anticipated, or calculated, area of fire spread, more sprinklers will operate than anticipated and more water will flow than the piping was designed to handle. As a result, pressure will drop, decreasing the individual sprinkler head discharge. With less pressure, the discharge pattern will not be within the design parameters of the sprinkler heads. The extinguishing efficiency of the sprinklers that are operating in the fire area will drop below 100% of design criteria, and continue to drop every time another head operates outside of the anticipated area of spread.

The wet pipe sprinkler system's strength lies in the fact that water is poised and ready to strike at the small fire, ready to shower its heat-absorbing, smothering, droplets of water directly on the heart of the potentially destructive combustion before the fire can intensify and spread.

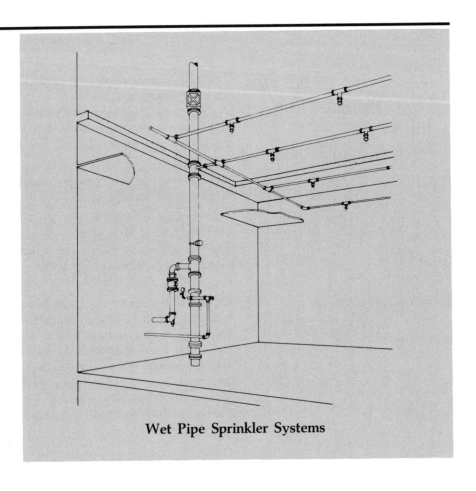

Wet Pipe Sprinkler Systems

Figure 3.1

Time is the greatest ally of a fire, which it needs as badly as it needs fuel and oxygen. The wet pipe system eliminates this time element by its speed in applying extinguishing water when the heat from the fire has reached the demand-for-attention point and actuated the sprinkler head directly over the fire.

Advantages and Disadvantages

Before examining the components that comprise a complete wet pipe system, the advantages and disadvantages of this system should be evaluated.

The major advantage of a wet pipe system is that piping is constantly filled with water under pressure for fast discharge on a fire. The disadvantage is that the system can only be used to protect areas where the temperature does not fall below the freezing point. Only when a sprinkler head is open and discharging is there any flow of water in the piping, which makes this system very susceptible to freezing.

The fact that the water in the piping has no movement, except when the system is in an operational condition, prevents corrosion from building up in the piping. Corrosion occurs when a constant or occasional water flow passes through piping. (Corrosion in other sprinkler piping is a saboteur working for the fire. It silently reduces the water passage area within the pipe by building up on the walls of the pipe, thereby cutting down on the amount of water that can pass through the pipe to the sprinkler head.)

In a wet pipe system, corrosion on the walls of the pipe is minimal since the volume of available corrosion chemicals in the water remains constant with no fresh supply of chemical available from flowing water. The temperature of the pipe and water assume identical levels in a relatively short time after initial installation and remain simultaneously constant. A change in room temperature between pipe and water will accelerate the deposit of scale or corrosion on the walls of pipe, but the stagnant water in a wet pipe system maintains a relatively constant pipe and water temperature, varying only if the room temperature changes.

Also, as corrosion builds up on the walls of the pipe, the smooth surface gives way to roughness which, in turn, creates small eddys and turbulence as the water flows through the pipe. This turbulence slows water down, decreasing velocity of the water measured in feet-per-second (FPS) and allows additional pipe-surface/water-surface contact time. A slow down of the water means more time for a pipe and water temperature differential to take effect, more time for chemical reactions to materialize, and more time for the corrosion process to have a maximum impact.

Wet Pipe System Alarm

When a sprinkler head is actuated and water discharges through the head, the sprinkler system has met two of the three required functions: detection and fire extinguishment or control. The third vital function is *alarm*.

A fire has been detected, and within the design parameters of the system, fire control has been initiated. As in any emergency, personnel are never sure whether or not their capabilities will meet the circumstances, and so they cry for help. In the case of the sprinkler system it may only be the need for the use of a hand extinguisher to snuff out the persistent fire under an obstruction that prevents sprinkler water from reaching the combustion zone. Or due to conditions that were not anticipated when the system was designed and installed, the sprinkler discharge can only hold an intense fire in check. In many cases, the professional assistance of firefighters with hose streams may be the only answer to complete extinguishment.

Water Damage

Consideration must also be given to water damage. With a wet pipe system the water discharge commences immediately upon head actuation, which may eliminate an incipient fire within a few minutes, but the water continues to flow. At the rate of 15 to 30 gallons per minute, a few minutes can mean a lot of water discharged and subsequently a great deal of damage to an occupancy.

Sprinkler system alarm and life safety are inseparable in every type of occupancy, but particularly where life safety takes precedence over property protection. The sound of a sprinkler alarm not only demands attention in the fire area, but also initiates emergency procedures throughout a facility. Where the lives of incapacitated or restrained individuals are at stake, all the responsibility cannot be delegated to the ability of the sprinkler system; the alarm it has sounded is merely an indication that there is serious trouble and that the system will do its best to provide the time necessary to perform orderly emergency procedures.

Initiating the Alarm

There are two basic methods for the initiation of an alarm when a fire has automatically changed a dormant wet pipe system into a dynamic fire fighter, and both of these methods use the most obvious feature, flow of water, in this status change. The two methods are water flow indicator and the alarm valve.

Water Flow Indicator

A water flow indicator of the paddle or vane type utilizes a flexible vane of thin metal or plastic (polyethylene). A hole is drilled in either the horizontal or vertical sprinkler pipe that supplies all of the sprinkler piping and sprinkler heads on the system. Then the flexible vane is rolled up and inserted into the hole. After the rolled vane is entirely through the hole it unrolls or expands to its original shape, completely filling the pipe with the exception of a small working clearance between the vane perimeter and the inside pipe circumference.

The stem of this vane is connected to a mechanical linkage which is encased in a cover. When there is a flow of water in the pipe the vane is deflected, and the vane-stem actuates mechanical linkage which, in turn, makes an electrical contact. The vane's flexibility prevents the obstruction of water flow in the pipe, and when the flow ceases the vane returns to its original horizontal position.

The size of the hole that is drilled in the sprinkler pipe to admit the vane is dependent upon the pipe size. For pipe sizes 3" and under, the hole diameter is 1-1/4" to 1-3/8" (depending on manufacturer), and for larger pipe sizes the hole is 1-7/8" to 2-1/8" in diameter (again depending on the manufacturer).

Clearance around the pipe is required to allow free movement of the vane or paddle.

When the unit consisting of vane and alarm switches is properly located with the vane in the pipe and expanded, the unit is bolted to the pipe with a "U" bolt and tightened to ensure a leakproof seal over the hole.

Any flow equal to one sprinkler head discharging will actuate the vane-type water flow detector. On some wet pipe systems this "one-head-flow-detection" is expanded to include not only the flow from any head on the system, but also to pinpoint which sprinkler line or which section of the sprinkler system is in an operating condition. In addition to the water flow indicator installed on the main supply pipe, water flow indicators may be installed on a cross main supplying a section of the building, or even on individual sprinkler branch lines supplying but a few sprinkler heads.

By installing these sectional or zoned water flow indicators, the exact area, or zone, where a sprinkler head has operated will be indicated on an alarm panel, thus eliminating the search for which area of the system a head has actuated and is flowing. In a large office building or institution containing a multitude of small rooms, or in a multi-story building, the time lost in locating the operating sprinkler head can be very costly to life safety, water damage, or fire control. Practical examples sometimes help in evaluation of such an explanation. For example:

> **Scenario 1:** Imagine that a six-story hospital with 50 patient rooms on each of 4 stories and the two remaining floors occupied by hospital support facilities such as operating rooms, storage rooms, and mechanical rooms, is completely sprinkled by a wet pipe sprinkler system with a vane-type flow indicator installed on the pipe which supplies the entire system. An alarm bell and light installed in the hospital telephone switchboard room and wired to the electrical contacts on the water flow indicator will be actuated by a flow of water from any sprinkler head anywhere on the entire system.

> It is 2 a.m. and the telephone operator is startled by the fire alarm bell. She is a trained and experienced individual who immediately contacts the fire department to report a "fire at Memorial Hospital!" But where does she direct the fire department when it arrives? There have not been any in-house calls to report smoke or water, and the security and maintenance crew that were the second parties to be alerted, have begun a floor-by-floor search but have not reported to the switchboard any positive results of their search. Time at this point is on the side of the fire condition.

> **Scenario 2:** Now consider the same hospital and the same 100% sprinkler protection with one significant difference. Each supply main that branches off the supply risers at every floor to supply the sprinklers on that individual floor are equipped with a water flow indicator. Now when the bell alerts the switchboard operator and a light on the annunciator is lighted behind the "Floor-5" window, she will call the fire department and report—"a fire, 5th floor, Memorial Hospital." The security and maintenance crews can be immediately sent to the 5th floor. Time is now on the side of the sprinkler system and the firefighters.

What has been described in Scenario 2 is a *local sprinkler alarm*. In Scenario 1, the bell alerts personnel on the premises, and someone calls the fire department. The human element is the weak link in this chain, and in a life safety situation, the need for professional help—the firefighters—is a vital element. With a local sprinkler alarm, the human link can be eliminated and the fire department summoned automatically when a sprinkler head operates, the flow of water begins, and the water flow indicator actuates. (See Figure 3.6 later in this chapter.)

The local alarm is not eliminated because the personnel at the facility must be alerted to a fire condition so they can take all available first aid firefighting measures and initiate an evacuation program, but an additional device is wired to the water flow detector alarm panel. An *alarm transmitter* which, when actuated, automatically transmits a signal over leased public telephone lines to a receiver in the fire station or fire alarm headquarters. This alarm receiver is marked for the facility it serves—"Memorial Hospital Fire Alarm" in this example. When this receiver is actuated, the fire department knows exactly where a sprinkler system is operating.

Every sprinkler system water supply is subject to water surges. This is a term which is self-explanatory. A wet pipe system is a tightly closed piping system. When the system installation is complete and the water is forced into the system, some air is trapped in the ends of the lines where it is compressed by the water until the pressure of this small amount of air equals the water pressure. When there is a sudden pressure surge on the supply main, such as an additional pump cutting in on the city water system, a large flow of water downstream of the sprinkler system on the city water main supplying the sprinkler system is suddenly stopped by the rapid closing of a valve. This excess pressure seeks the area of least resistance, and since water is not compressible, the excess water pressure further compresses the trapped air causing a momentary flow of water in the sprinkler piping. This momentary flow will deflect the water flow indicator vane, operate the mechanical linkage, and the electrical alarm switch makes contact, and activates the alarm system. Since these water surges can occur with some frequency, especially on a city water system, this could mean many fire alerts and with a remote alarm system the frequent rolling of the municipal fire equipment. False alarms can be more dangerous than no alarm at all, since the old "cry wolf" can destroy the time advantage by waiting to check and see if it is a false alarm before alerting personnel and the fire department.

To control this unavoidable danger, water flow indicators are equipped with a *retard feature*. This retard feature literally retards the transmission of the electrical signal to the alarm system, and although it is factory-set, it can be adjusted in the field after installation. The retard element can usually be adjusted for a 0 to 70 second delay, which means the alarm signal can be transmitted the instant the vane is deflected, or the alarm signal transmission can be delayed for up to 70 seconds after the vane is deflected by a water flow.

This retard must have an instant recycling feature so that a sequence of flows, each of less duration than the predetermined retard period, will not have a cumulative effect. For example:

> A 30-second retard is actuated by a water pressure surge. Fifteen seconds into the retard, another surge occurs and now there is an accumulative retard of 45 seconds. Fifteen seconds later, another pressure surge occurs and now there is an accumulative retard of 75 seconds. With the instant recycling feature, the first retard time remaining would have been canceled when the second surge occurred causing a new 30 second retard, and so on with the subsequent surges.

Adjusting the retard must be accomplished by experienced personnel with the approval of the authorities having jurisdiction, but the instant recycling feature must always be specified when a flow switch is part of the specification.

Alarm Valve

The water flow indicator functions electrically. Therefore, if there is a power failure and the facility does not have a back-up power system or the alarm system is equipped with a battery back-up, a flow of water in the sprinkler system will not produce an alarm. Primarily through the influence of insuring interests, with their concern for property protection and their fear of excessive water damage from undetected sprinkler discharge, most wet pipe systems are equipped with an *alarm valve*. (See Figure 3.2.)

The alarm valve, often referred to as a glorified check valve, is a special check valve with water flow detecting capabilities. This device not only detects flow, but can operate a mechanical local alarm, a water motor alarm, or activate an electrical alarm device.

The alarm valve is installed in the main sprinkler supply line close to the point where the underground water supply enters the building. The alarm valve is installed in the pipe that supplies water to the entire sprinkler system so that a flow of water from any operating sprinkler head on the system must pass through the alarm valve.

All alarm valves are different in some respect, but the basic design and function of all alarm valves is identical. The alarm valve body is cast iron with standard flanged or grooved connections. It contains a clapper similar to a check valve, with a rubber or neoprene seat where it makes contact with the bronze water passage opening.

Wet Pipe Sprinkler System Alarm Valve

(courtesy of the Viking Corporation)

Figure 3.2

Being a wet pipe system, the piping and alarm valve are completely filled with water, and the pressure will be equal on the top and bottom, supply and system side of the clapper position. If water surges have occurred which force excess pressure through the alarm valve compressing the air in the sprinkler line piping, the clapper will open momentarily to admit the excess pressure, and the close again when the surge recedes, trapping the higher pressure on the system side of the clapper. For this reason there are two different pressure gauge readings on an alarm valve. One gauge monitors water pressure on the supply side of the alarm valve clapper, and the other monitors rate pressure on the system side of the alarm valve clapper. This difference is not objectionable since it holds the clapper in the closed position and prevents the possibility of false alarms. This will become clear in further examination of the alarm valve function. When a sprinkler head opens and allows water to discharge, there is a steady flow of water through the alarm valve which raises the clapper. While the clapper is in the closed position, it seals off an opening to the alarm line. When the clapper raises, the water pressure forces water into this small alarm line pipe opening. This feature separates the alarm valve from the standard check valve, and justifies calling this piece of equipment an alarm check valve.

A slight amount of excess pressure on the system side of the alarm valve clapper diminishes the chance that the clapper will raise with a surge of supply pressure and expose the alarm line opening. This momentary raising of the clapper will occur from time to time, and water will be admitted to the alarm line, so provisions must be made to prevent false alarms. Like the water flow indicator, alarm actuation must be retarded to prevent false alarms.

Water Motor Alarm

The water that enters the alarm line, although a small stream, has the same pressure as the sprinkler system water supply, and this pressure is used to operate a mechanical water motor alarm. This alarm device consists of a paddle wheel in a sealed housing. The pressurized water stream from the alarm line discharges against the fins or paddles of the water wheel and mechanically turns the center shaft on which the paddle wheel is mounted. On the other end of this shaft is mounted a striker which when rotated within a metal bell casing continuously strikes, or rings the bell.

The water motor, or paddle wheel, is usually installed on the inside wall of a building with the shaft extending through the wall in a shaft sleeve, and the bell mounted on the exterior side of the wall.

This is strictly a local alarm for the wet pipe system, and the efficiency of the water motor as an alarm is only as great as its sound range, and personnel within that sound range. It is a back-up, a simple, reliable, mechanical, water flow indicating alarm. If it is required, or desired, to supplement this local alarm with an electrically operated water flow alarm, a device known as a pressure switch is also used in conjunction with, or in lieu of the water motor alarm, if approved by the authorities having jurisdiction.

Pressure Switch

The pressure switch on an alarm valve is also activated by water pressure transmitted through the alarm line. This pressure acts on a diaphragm in the pressure switch, forcing it to move and thus make electrical contact. The details of the pressure switch will be further described in the section on alarm devices.

Retard Chamber

System water pressure transmitted through the alarm line is the force which operates both the mechanical and electrical alarm devices on a wet pipe sprinkler system. The next device which warrants consideration is the retard chamber which prevents water surges from producing false alarms.

The alarm line runs from the alarm valve to the retard chamber and from the outlet on the retard chamber to the water motor alarm and/or the electric pressure switch.

The retard chamber is merely a holding tank, having a capacity of approximately two gallons. The inlet to the retard chamber is through a restricted orifice approximately 5/16″ in diameter attached to the 3/4″ alarm line from the alarm valve. This orifice restricts or slows down the volume of water which the water pressure in the alarm line can force through the opening and into the retard chamber. In other words, it restricts the flow of water to the retard chamber. (See Figure 3.3.)

The retard chamber must completely fill with water before it can discharge through a 3/4″ opening at the top of the chamber. Water continues on through the alarm line to actuate the water motor and/or force the diaphragm in the presence switch to deflect. This makes an electrical contact to complete a circuit, light lights, ring bells, or actuate a transmitter to automatically conduct a signal to a remote alarm. When the wet pipe sprinkler system is subject to abnormally large water surges, it may be necessary to install two retard chambers in tandem. The 3/4″ outlet at the top of the primary retard chamber is piped into the 3/4″ drain opening in the bottom of the secondary retard chamber, and the outlet to the alarm devices is connected to the alarm line at the top of the secondary chamber. Both retard chambers must completely fill with water before the water pressure can be transmitted to the alarm devices.

At the base or bottom of the retard chamber is a small 1/8″ restricted orifice drilled into a 3/4″ drain line. If a water surge has passed water into the alarm line and thence to the retard chamber, and has half filled the chamber by the time the alarm valve clapper resets and shuts off the alarm water, the water in the retard chamber will drain out slowly through the 1/8″ restricted orifice into the 3/4″ drain pipe (which is usually connected to the main drain pipe from the alarm valve).

There are retard chambers with more sophisticated entry and drain mechanisms. The retard chamber with a clapper and mechanical linkage arrangement is one example. When water is introduced into the retard chamber from the alarm line, it forces a small clapper to open the inlet waterway, and the clapper (being connected by an arm to a diaphragm) automatically closes the drain opening. When the water pressure on the alarm line is relieved by the alarm valve clapper resetting and closing the alarm line outlet, the retard chamber inlet clapper resets due to a spring mechanism and subsequently releases pressure on the drain diaphragm. This, in turn, opens the drain opening and the false alarm water in the retard chamber drains off into the drain line.

If the alarm system consists of a retard chamber and pressure switch only, it will be necessary to bleed-off the air in the retard chamber since the pressure switch is connected to the 3/4″ outlet from the top of the retard chamber. The entire complex, with water covering the chamber drain opening, is air-tight. As the water rises in the chamber, the air will be compressed until it reaches the pressure of the water. At this point of equalization no more water will enter the chamber. The pressure switch

Pressire Switch
(Order Separately)

2" x 3/4" Pipe Bushing

1/2" x 3-1/2" Nipple

1/2" Plug

Outlet

1/2" x 1/4" x 1/2" Tee

3/4" x 1/2" x 1/2" Tee

3/4" x 2" Nipple

1/8" Restriction

1/4" NPT (Female) To open
Drain or Outlet of System
Main Drain Below Retarding
Chamber Drain Connection.

2" x 1/2" Pipe Bushing

Inlet

1/2" Inlet

MODEL C-1
RETARDING
CHAMBER

Figure A. Cutaway of Model C-1
Retarding Chamber

Figure B. Circuit Closer Vent Trim

Retarding Chamber

(courtesy of the Viking Corporation)

Figure 3.3

has a very tiny air-compensating vent which, under normal conditions, maintains an equal air pressure on both sides of the electric contact diaphragm. This compensating vent may not bleed off enough air pressure to prevent this air/water pressure equalization, and this can very possibly prevent the diaphragm from receiving sufficient alarm water pressure to deflect sufficiently to make the electric alarm contact. The water pressure acting on the diaphragm is necessary to cause sufficient movement for a positive electrical contact. To alleviate this condition, a line containing a restricted orifice is installed between the retard chamber and pressure switch. This line discharges into the main alarm valve drain line and bleeds off the excess air pressure. If this bleed-off arrangement is not provided, and a water motor alarm (which bleeds-off pressure) is not installed, an automatic air vent on the retard chamber must be installed.

Alarm Valve Trim

The standard trim for an alarm valve consists of two pressure gauges to monitor supply and system water pressure, and each gauge is equipped with a small (usually 1/4") control valve so that a deflective gauge may be replaced without disturbing the entire system. Each pressure gauge line has a 1/4" plug which can be removed, providing an opening for an inspector to install his pressure gauge to check the accuracy of the installed gauges.

Taking suction from the supply side of the alarm valve clapper is a small alarm test line piped through a bypass test valve, which is normally closed and only used for testing the alarm system. It is then piped to the retard chamber. The inspector can open this bypass test valve and actuate the alarm system with a simulated flow from the alarm line. The retard chamber can also be removed for repair or replacement without shutting off the system water supply by installing an *indicating control valve*, which visually indicates the open and closed position between the alarm line opening under the alarm valve clapper and the retard chamber.

The *wet pipe system main drain* is connected to an opening in the alarm valve body above the clapper, and is piped through an angle control valve to discharge through the building wall to the atmosphere or to an open floor drain, roof drain, or other discharge receptacle that can accept the discharge flow. The main drain can be of more importance than merely draining off the system water when it becomes necessary to make repairs to the system. The same holds true when the area protected by the system will be subject to freezing temperatures and the water must be drained out of the system. If a sprinkler head should be damaged (hit by a lift truck, a rolling scaffold, a ladder, etc.) and commence discharging water, the first action taken to avoid serious water damage is to close the main control valve to shut off all water supply to the system. If the main drain is simultaneously opened wide, a great deal of system water can be drained off through the main drain instead of running off through the open sprinkler head. This procedure will immediately relieve some of the back pressure which is forcing water out of the open-head even after the main water supply is shut off. This is one reason why the main drain discharge must be unrestricted, and must discharge where the full flow of the drain line will not cause additional water damage.

The size of the drain line is dependent upon the size of the system. The size of the system riser supplying the system determines the system size: A 4" or larger system requires a 2" drain, a 2-1/2"-3" riser needs a 1-1/4" drain, and smaller risers require a 3/4" drain.

Fire Department Pumper Connection

The fire department pumper connection, or siamese connection, is a common sight on the exterior wall of a building. The Y-shaped pumper connection has caps covering the two openings of the 2-1/2″ pipes which protrude from the casting that is the body of the Y. (See Figure 3.4 for an illustration of the four-way pumper connection.) On buildings where appearance is a factor, all that need be seen are two, usually chrome-plated, caps installed on concealed 2-1/2″ pipes projecting through an oblong metal wall plate. These wall type fire department pumper connections are commonly sized 2-1/2″ by 2-1/2″ by 4″; meaning that the two 2-1/2″ outlets are "siamesed" into a 4″ casting into which is screwed a 4″ supply pipe.

This 4″ pipe is piped into the main sprinkler riser—the pipe that supplies all the water to every head on the system. (The diameter is used for example only. Under certain conditions, a 6″ or 3″ pipe may be used.) As soon as the fire department arrives at the scene of a fire, they will connect a line from their pumper truck to the closest hydrant. From two pumper truck discharge outlets, two 2-1/2″ lines will be connected to the two 2-1/2″ outlets on the sprinkler system siamese connection. The fire department can now pump water at a high pressure and volume directly

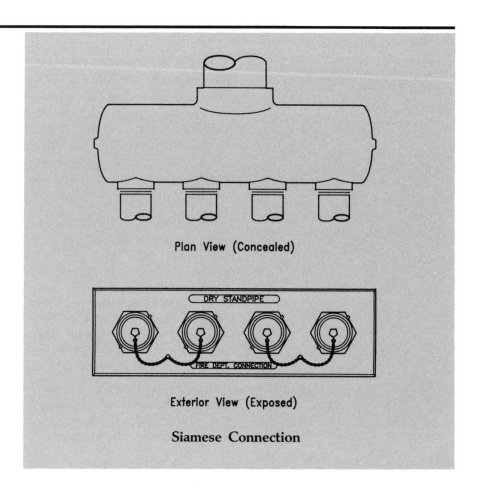

Plan View (Concealed)

Exterior View (Exposed)

Siamese Connection

Figure 3.4

into the sprinkler system. If the fire has overpowered the sprinkler system (more heads have opened than were anticipated in the hydraulic design of the system) this water being pumped into the system can augment the volume and increase the pressure to overcome the increased demand of the discharging sprinklers. (See Figure 3.5.)

The records of the National Fire Protection Association (NFPA) are a tragic reminder of how often a main water supply control valve to a sprinkler system is not reopened after a system repair is completed. In these cases, the fire department pumper connection is the only source of water supply to the firefighting sprinkler heads. For this reason, the fire department pumper connection must always be connected to a sprinkler riser on the system side of all control valves.

The pipe from the siamese connection to the sprinkler system must always have a check valve through which the fire department can pump water into the system, but when not in use, the system water pressure will not leak out, or flow out through the siamese connection when caps are removed and hoses connected. In cold climates where the siamese connection on the exterior wall of the building is subject to freezing, the check valve keeps the line to the siamese dry and empty of water.

To drain off water in this section of line after the siamese connection has been used, to drain off any leakage of water through the check valve, and to drain off condensation, *an automatic ball drip* drain is installed on the siamese side of the check valve either at the check valve or at a low point on this segment of piping. One type of automatic ball drip consists of a small, usually 1/2", cast iron body drain containing a solid metal ball

Fire Department Pumper Connection

Figure 3.5

which rests in a depression beside the open drain opening. When the ball drip drain is under pressure while the siamese is in use, the water pressure forces the ball over the drain opening thus closing off the drain.

One additional element must be considered. Suppose the sprinkler system water supply has a pressure of 60 pounds per square inch (psi) and is fed from a city water main. If the fire department pumps into the siamese connection at 100 psi with a capacity of 500 gallons per minute (gpm), and the sprinkler heads that have opened are only discharging a total of 200 gpm, the water will seek the path of least resistance, the 60 psi water supply. This will force the supply water back into the city water mains, completely nullifying the fire department siamese water and pressure to the open heads on the system.

Every sprinkler system must have the auxiliary water supply afforded by the siamese connection. In addition, every fire department siamese connection inlet pipe must have a check valve to prevent system water discharge back through the siamese. Every sprinkler system must have a means of checking the water supply piping to prevent flow back to the supply source when the fire department is pumping into the siamese.

When a vane or paddle type water flow indicator is the alarm device on a wet pipe system, a check valve must be installed in the main sprinkler water supply on the water supply side of the connection of the siamese. This is connected to the sprinkler riser to check the flow of the siamese back into the supply.

When an alarm valve is the water flow alarm device on a wet pipe system, the check valve feature of the alarm valve can be used to accomplish this supply check. The siamese supply is connected to the sprinkler system on the system side of the alarm check valve.

The siamese connection on the exterior wall of the building should be located not less than 18″ nor more than 5′ above grade (or the access level of the fire department).

Main Control Valve

The main control valve is installed in the main water supply pipe to the sprinkler system, and controls all of the water to the sprinkler system. The control valve may be installed inside the building where the underground supply pipe enters the building, or it may be installed outside the building in the underground supply pipe. (See Figure 3.6.)

Inside or outside, all sprinkler systems, and in fact all fire protection control valves, are required to have one feature—visual indication of an "open" or "shut" position. This section will consider only the inside, or above ground, control valve.

The terms "inside" or "outside" control valves refer to valves installed in the underground pipe outside the building, as opposed to valves installed in the steel pipe of the sprinkler system above ground, and inside the building. A commonly used inside control valve for sprinkler systems is the outside screw & yoke (OS&Y) valve.

The OS&Y valve is sometimes referred to as a rising stem gate valve because when open, the stem extends far above the hand wheel, and when closed the stem is flush with the hand wheel. This meets the requirement for all fire protection valves controlling water supplies; namely, a visual inspection will tell at a glance whether the valve is open or closed. As a result, all valves controlling water supplies must be indicating type valves, which means they must visually indicate whether they are open or shut.

There are several indicating type control valves on the market for use on fire protection systems. Except when the use of an OS&Y valve is required by NFPA or the insurance carrier, the use of these space saving butterfly control valves is rapidly replacing the OS&Y valve.

The control valve on a sprinkler system serves several basic functions. This is necessary in order to shut off the water supply to the system so that the system can be repaired, altered, or increased in size.

When the sprinkler system has suffered physical damage, and is discharging water through a broken pipe, fitting, or sprinkler head, the supply of water to the entire system can be stopped by closing the main control valve.

When the sprinkler system has performed within the efficiency of the design parameters and extinguished a fire, the water damage can be kept to a minimum by closing the main control valve and stopping the water supply to the heads that have been opened by the fire. At this point, it would be inexcusable not to mention the tragic record of major fire losses that have resulted from this misuse of the main control valve. A fire that produces enough heat to operate a sprinkler head cannot be safely declared officially extinguished by the average layman. All too often in their haste to avoid water damage, a main control valve will be prematurely closed because the fire appeared to be out. Control valves are not always conveniently located within eyesight of the fire. By the time the individual who has been dispatched to the basement where the valve is located returns to the third floor where the fire has occurred, he finds that the fire has re-ignited and is using the time for his return trip to reopen the valve to feed itself, uninterrupted by the sprinkler discharge. The time element involved has changed the complexion of the fire situation from a controlled to an uncontrollable fire condition.

Control Valve

Figure 3.6

Many NFPA fire records have shown that a control valve was shut off too quickly, and when the error was discovered the fire had made it impossible for an individual to approach the valve location to reopen the control valve.

Without a water supply, a sprinkler system is nothing more than a complex of pipe and fittings, which points out another facet of the importance of the open control valve. NFPA fire records indicate that automatic sprinklers have a history of 96.2% satisfactory performance. The majority of the 3.8% unsatisfactory performance record is accountable to a main control valve closed for sprinkler system maintenance or alterations and, through neglect or human failure, not reopened. In the section on alarms and supervisory systems, we will examine the available means for eliminating this costly and dangerous situation.

The fire department uses the sprinkler system control valve in situations where, in their professional opinion, the fire has overpowered the sprinkler system. So many heads have opened, and so much water is being discharged, that it is affecting the water supply and pressure at the fire hydrants. The sprinkler system has lost its value at this point, and has in turn become a deterrent by robbing the hydrant hose streams of vital volume and pressure, and the fire department will close the control valve.

Another case where the control valve is shut occurs when there is a flash fire: When an ignition source ignites fumes which have spread over the entire floor area, a massive flammable liquid spill covers a wide area, or the preheated combustible construction of an entire room. The flash of heat can open heads far removed from the primary fire source, and is usually intense enough to fuse sprinkler heads, but not of sufficient duration to sustain continuing combustion within its area of influence. The result will be a hundred sprinklers operating on a system calculated for the operation of 30 sprinklers, and the drain on the water supply, both pressure and volume, far exceeds design parameters.

The fire department will also close the main control valve to save water for their hose streams when a sprinkler system supply pipe has been broken by an explosion, collapsing wall, or structural member.

The location of a main control valve on a wet pipe system incorporates what we have learned about the uses of a main control valve and the fire department pumper connection. The control valve on a wet pipe sprinkler system is always installed on the supply side of the fire department pumper connection. By following this rule, a closed control valve can never prevent or obstruct the use of the fire department pumper connection.

Inspector's Test

The wet pipe sprinkler system has been installed, and by all appearances is ready to perform its designed function. Unlike almost every other building system, however, its operational ability will not be demonstrated on a daily basis. The sprinkler system may not be called upon to function for 10 or 20 years, but when it is needed, the system must function without fail. The eruption of fire is no time to discover a mechanical defect that affects the system operation, and sprinkler systems are seldom afforded a second chance.

In order to test this completely closed and sealed system without the expensive and impractical fusing of a sprinkler head, each wet pipe system is equipped with an inspector's test connection. (See Figure 3.7.)

The inspector's test connection consists of a 1" valve installed in a 1" pipe that takes supply from the wet pipe sprinkler system. When this valve is opened, water is discharged through a smooth bore corrosion-resistant outlet or bushing that produces the flow equivalent to the flow from one open sprinkler head. Opening the inspector's test valve will, if the system is designed, installed, and maintained properly, operate all of the wet pipe system equipment and actuate all alarm equipment. In order to obtain a true reading on the system, the inspector's test should be installed at the end of the branch line that is most remote from the system supply, and at the highest point on the system. If the supply pressure is not sufficient to provide adequate volume and pressure, the inspector's test will not have sufficient pressure to discharge an adequate amount of water to operate the system or alarm equipment. The inspector's test is a valuable tool for the experienced inspector.

Wet-Pipe System Test Connection

Figure 3.7

Discharge

The discharge from the inspector's test must be unobstructed and visible. Discharge may be either to atmosphere or, where this is not practical, the test pipe may terminate in a roof or non-sanitary drain pipe capable of accepting full flow under system pressure. A sight test connection containing a smooth bore corrosion-resistant orifice providing a flow equivalent to the flow from one sprinkler is installed between the inspector's test valve and the connection to the drain line.

Location

The test valve should be located where it will be readily accessible and not over seven feet above the floor. In the discussion of water flow indicators, the use of these devices at various locations on a wet pipe sprinkler system to afford pinpoint location of a water flow requires the installation of an inspector's test connection to facilitate testing each of these water flow detection devices.

Where the test connection to operate a vane-type flow switch cannot discharge to the exterior of a building and there are no suitable plumbing drains to accept the test discharge, a closed system test and drain may be used. For example, a multi-story dormitory completely protected with a wet pipe sprinkler system:

> In order to have an alarm panel that indicates which floor has a sprinkler head flowing, the panel is wired to a flow switch on each floor, installed in the pipe supplying the sprinkler on the floor.

> The supply to the sprinkler on each floor of the dormitory is controlled by an indicating control valve. Should a leak have to be fixed, or a sprinkler head replaced, the floor control valve can be closed, isolating that one floor sprinkler system and leaving all other floors under sprinkler protection. In order to be sure that the water flow switch on each floor is working properly, and to drain the piping once the control valve is closed, a test and drain set-up can be installed as illustrated on each floor. (See Figure 3.7.)

The test and drain set-ups discharge into a dedicated drain riser, which picks up the test and/or drain water at each floor and carries it to an acceptable location at ground floor level for discharge.

Wet Pipe System Drainage

References have been made to shut the main control valve and drain the system for maintenance, alterations, or to freezing temperatures. The ideal design of a wet pipe sprinkler system would be to have all of the water in the piping network drain back to the main drain, but due to changes in elevation, there are usually low points, trapped sections of piping, or heads on the system requiring auxiliary drains. To quote from the NFPA Standard 13 for the Installation of Sprinkler Systems, "all sprinkler pipe and fittings shall be so installed that the system may be drained."

Until recently, the requirement to accomplish wet pipe sprinkler system drainage was to pitch the pipe so that all water, exclusive of trapped areas, would automatically drain back to the main drain. This rule has since been changed so that now wet pipe sprinkler piping may be installed level. It was determined that the inherent pipe size change in sprinkler piping with a level center-line would cause the water to naturally flow from the small to the large diameter pipe, and thence back to the main drain. This change has been a tremendous sprinkler design advantage where long runs of pipe are coupled with close clearance tolerances, such as above a suspended ceiling.

The requirements for draining trapped sections of a wet pipe sprinkler system are based on the number of heads that are trapped or the number of gallons of water that will drain back to the main drain by gravity. These auxiliary drains have no bearing on the fire protection efficiency of the system, and are installed to minimize water damage in draining the system.

A drain plug in a small diameter pipe creates the impression that only a small amount of water is trapped in the piping. Sometimes due to construction or occupancy, sprinkler piping is not readily visible. It does not take much imagination to visualize what happens when a workman removes a 1″ plug and holds up a 5-gallon bucket to catch the water, not knowing that the 1″ pipe is on the end of a large trapped section of sprinkler piping containing 30 gallons of water.

Where it is obvious that the trapped section contains less than five gallons, a drain plug is all that is required. However, when a trapped section appears to contain several gallons, supplies quite a few sprinkler heads, or has a rather large pipe supply, the installation of a 1″ drain valve will facilitate draining. The ideal situation is to discharge a drain line to the exterior of the building. When this is possible, but the exterior ambient temperature is subject to freezing (32°F or less) at least 4′ of pipe should be installed beyond the drain valve in a warm room.

Dry Pipe Sprinkler Systems

The wet pipe sprinkler system, because it is simple, efficient, and economical to install and maintain, accounts for about 75% of all sprinkler installations. Dry pipe sprinkler systems, on the other hand, account for a much smaller percentage of sprinkler systems in use. They are designed for use in facilities in which it is either impractical or impossible to provide the heated environment necessary for the proper functioning of a wet pipe system. The dry pipe system provides fire protection to unheated structures, and in some cases, specific unheated areas of the structure in which the system piping is subjected to freezing temperatures. Use of a traditional wet pipe system in an unheated zone would result in frozen pipes, which could block the flow of water to a fire or cause considerable water damage should the pipes burst. This section examines the basic operation of the dry pipe sprinkler system, including advantages and disadvantages of the system and an explanation of the system's components. (See Figure 3.8.)

Basic System Operation

The dry pipe system, like the wet pipe system, uses water as the extinguishing agent. Unlike the wet pipe system, however, the dry system piping is filled with pressurized air or nitrogen. When a sprinkler head fuses, the air or nitrogen pressure is dissipated through the open head. The piping then fills with water and discharges through the open head.

Understanding how a dry pipe system operates requires a brief lesson in physics. A moderate amount of pressure on one side of a large surface can equalize a greater amount of pressure on a small surface on the opposite side. The combined pressure on the total surface area can be found by multiplying the amount of pressure expressed in pounds per square inch (psi) by the number of square inches.

51

Most dry pipe valves are designed so that a moderate amount of air pressure on the larger diameter "air side" of the clapper in the dry pipe valve will hold back a much higher water pressure acting on the smaller surface "water side" of the clapper. (See Figure 3.9.) When the air pressure can no longer contain the water pressure, the valve will open by raising the clapper. The difference between the air pressure and the water pressure at this point is expressed as a ratio. This ratio is called the *differential*. Dry valves that operate on this principle are called *differential-type dry pipe valves.*

Basically, dry pipe sprinkler system piping is filled with pressurized air or nitrogen. When a sprinkler head operates, the compressed air is discharged faster than the system's compressed air source can replenish it. The pressure on the air side of the clapper drops to the valve "trip point," which is the point at which the water pressure overcomes the air pressure on the air side of the clapper. The clapper flies open and water rushes through the system to the open sprinkler head(s).

Advantages and Disadvantages of Dry Pipe Systems

The basic advantage of a dry pipe system is that it can be installed in areas subject to freezing. Dry pipe sprinkler systems are the only way to economically provide sprinkler protection in unheated areas. Unfortunately, there are drawbacks to this system.

Dry Pipe Sprinkler Systems

Figure 3.8

One drawback already mentioned is the delay that occurs between the time the head fuses and the time when water is discharged. This is time that enables the fire to intensify and spread.

Another disadvantage directly related to the first is that as the area of a fire increases, so does the number of heads that will open in that area, called the *area of application*. NFPA records indicate that on the average, more sprinklers are actuated in areas protected by a dry pipe system, than in areas protected by a wet pipe system. In fact, the area of application for a wet pipe system is increased 30% when a dry pipe system is used.

Another disadvantage of this design is the time lag inherent in the operational sequence of a dry pipe system. When the sprinkler head is actuated by the heat of a fire, and the sprinkler head discharge orifice is opened, it is necessary to reduce the air pressure that fills the system to the trip point by releasing this air through the sprinkler orifice. It is not unusual to have a dry pipe system containing 300 or 400 gallons of air at 30 or 40 pounds of pressure. With standard 1/2″ orifice sprinkler heads, this can mean a considerable time delay to evacuate a sufficient volume of air through the orifice to reduce the pressure to the trip point. Once the water is released into the system, it must overcome the back pressure of the remaining air in the piping in traveling to the open sprinkler head. While sprinkler head actuation-to-water discharge time in a wet pipe system is instantaneous, with a dry pipe system it may take 60 seconds or longer, depending on the size of the system and the air pressure level. This is precious time that could be used extinguishing the fire.

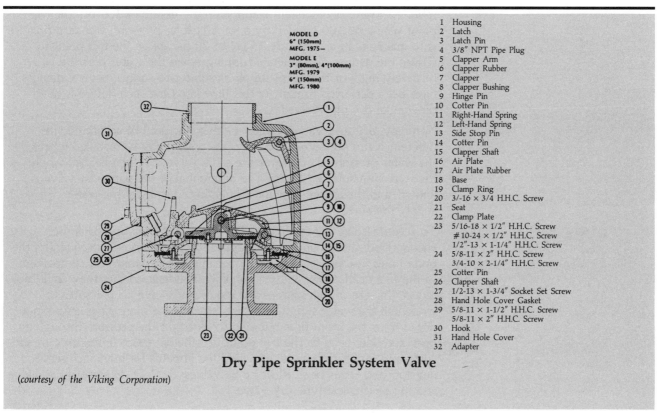

MODEL D
6″ (150mm)
MFG. 1975—
MODEL E
3″ (80mm), 4″(100mm)
MFG. 1979
6″ (150mm)
MFG. 1980

1 Housing
2 Latch
3 Latch Pin
4 3/8″ NPT Pipe Plug
5 Clapper Arm
6 Clapper Rubber
7 Clapper
8 Clapper Bushing
9 Hinge Pin
10 Cotter Pin
11 Right-Hand Spring
12 Left-Hand Spring
13 Side Stop Pin
14 Cotter Pin
15 Clapper Shaft
16 Air Plate
17 Air Plate Rubber
18 Base
19 Clamp Ring
20 3/-16 × 3/4 H.H.C. Screw
21 Seat
22 Clamp Plate
23 5/16-18 × 1/2″ H.H.C. Screw
   #10-24 × 1/2″ H.H.C. Screw
   1/2″-13 × 1-1/4″ H.H.C. Screw
24 5/8-11 × 2″ H.H.C. Screw
   3/4-10 × 2-1/4″ H.H.C. Screw
25 Cotter Pin
26 Clapper Shaft
27 1/2-13 × 1-3/4″ Socket Set Screw
28 Hand Hole Cover Gasket
29 5/8-11 × 1-1/2″ H.H.C. Screw
   5/8-11 × 2″ H.H.C. Screw
30 Hook
31 Hand Hole Cover
32 Adapter

Dry Pipe Sprinkler System Valve

(courtesy of the Viking Corporation)

Figure 3.9

Another disadvantage of the dry pipe system is *corrosion*. The piping is filled with compressed air supplied to the system from one of two sources—air compressor or shop air.

Both of these air supply methods obtain their air supply from the atmosphere, and at ambient temperature. The air supply may not be the same temperature as the piping of the dry system, and if relatively warm air is pumped into dry system piping installed in a cold area, condensation of the moisture in the warm air will collect on the interior walls of the system piping. This creates an ideal atmosphere for the formation of corrosion or scale in the piping.

Aside from producing corrosion, condensation causes another problem. Where the temperature differential is severe (for example, where a dry system is installed in a walk-in freezer and the dry valve and air compressor are located outside of the freezer) the small piping fills with water, freezes, and can crack the pipe. For this reason, the system air supply to the compressor must maintain a temperature closely approximating the temperature of the area where the piping is installed.

Condensation and corrosion can be fairly well contained, except under severe conditions. For example, a dry pipe system protecting an unheated warehouse where the temperature of the piping under a metal roof may reach 100°F in the daytime, and drop to 60°F at night will be particularly vulnerable. It is almost impossible to control the air temperature being supplied to the system to maintain required pressures.

Differential Dry Pipe Valves

The differential-type dry pipe valve has a trip ratio of approximately 6:1. In other words, the valve operates when the air pressure in the system is reduced to approximately one-sixth of the system water supply pressure. For example, in a differential dry pipe valve with a 6:1 ratio, operating on a system water supply pressure of 100 psi, the trip point would be 100 divided by 6, or 17 psi air pressure. Standard practice is to maintain an air pressure 15 psi to 20 psi above the trip point to reduce the danger of accidental tripping when the water pressure is relatively high, or the water supply is subject to surges. Several devices have been perfected to narrow this time lag. One such device is the differential dry pipe valve.

Naturally, dry pipe valves cannot be manufactured to accommodate different water pressures. A 6" dry pipe valve may be installed where the water pressure is 60 psi, or the same valve may be installed where the water pressure is 125 psi. The differential of the particular valve is known, and the air pressure must be adjusted to meet the water supply pressure.

To accommodate the relatively high water pressures that could be encountered, the air side of the clapper has to be large. Subsequently, the body housing these large clappers became rather massive in comparison to the size of the alarm valve. Manufacturers realized that they could not reduce the size of this clapper and require excessive air pressures to overcome the water supply pressure, because the more air pressure in a dry system, the more time required to bleed off the pressure through an open sprinkler head to the trip point. Aesthetics, space limitations, freight costs, and installation costs, provided the impetus for manufacturers to develop two types of small dry pipe valves: the low differential dry valve and the mechanical type dry valve.

Low Differential Dry Valve

The low differential dry pipe valve has a differential in the range of 1.0 and 1.2 to 1 instead of the 5: or 6:1 ratio common with the ordinary differential dry pipe valve. The low differential dry pipe valve resembles an alarm check valve both in size and operation. The air pressure on the sprinkler system side of the clapper (the air side) actually exceeds the pressure on the water side of the clapper. The valve operates when the air pressure is reduced to only 10% less than the water pressure. With a water pressure of 100 psi, for example, the air pressure would be 115 or 120 psi. To operate the low differential dry pipe valve, the air pressure would only have to drop to approximately 10% below the water pressure (in this instance 90 psi). With air pressure this high, bleeding off 20 or 30 psi through an open sprinkler orifice would require only a few seconds, unlike the excessive trip time of ordinary differential dry pipe valves.

The low differential dry pipe valve was found to have an advantage beyond its smaller size. An ordinary dry pipe valve provides an unobstructed flow of water when the valve is tripped. With the conventional 5: or 6:1 differential-type dry valve, using low system air pressures as compared to the supply water pressures, the rush of water into the system is tremendous, since the low air pressure offers little compressive resistance. In the low differential-type dry pipe valve, on the other hand, the system air pressure at trip point nearly approximates the supply water pressure, so that the velocity of water entering the system is greatly reduced. Aside from reducing the possible damage to the sprinkler system caused by a pressure surge of high-velocity water through a tripped dry valve, the amount of debris, corrosion, scale, and sediment entering the system is reduced with a low-differential dry pipe valve because the carrying force of the low-velocity water is greatly diminished. This is crucial to the effectiveness of the system, since scale and sediment carried to the open heads can clog the sprinkler head orifices and impede water discharge during a fire.

Note how similar the low differential dry pipe valve is to the alarm valve, not only in appearance, but in the method of charging the alarm line. A pilot valve on the clapper arm closes the alarm line opening when the clapper is in the closed position. When the valve is tripped, the clapper is raised and the alarm line orifice is opened. Note, too, that like the alarm valve, the clapper is not latched in the open position, but is free to swing shut.

It is not accepted practice to use this valve as a check valve opposing the fire department pumper water pressure because of the possibility of damaging the rubber clapper ring that seals the clapper. As a seal, this clapper is important regardless of the type of valve. The fire department pumper connection to the dry system must be made on the wet side of the dry pipe valve due to the inability of the check valve in the fire department pumper line to hold air pressure, and the inability of the dry valve to act as a check valve.

In the dry pipe valve, special rubber valve rings are used on clappers to maintain a tight seal when the clappers are in the closed position, and to compensate for any minute metal pocket, pitting, or minor metal-to-metal dimensional variations that could allow air to escape. As a further safety factor, the air side of the clapper is primed with enough water to completely cover, and thereby seal off, the clapper, further preventing air loss. This priming water also keeps the rubber seals from drying out and shrinking or cracking.

Mechanical-Type Dry Valve

The mechanical-type dry pipe valve also permits the use of a smaller body to house the functional mechanism of the valve. The air pressure in a dry system controlled by a mechanical-type dry valve acts on a small disk, diaphragm, or clapper which, in turn, is arranged to actuate levers, links, and latches to hold the main clapper in the closed position and to release the main clapper when the air pressure is relieved.

There are as many types of mechanical valves as there are valve manufacturers. The maintenance of these valves is in direct proportion to their mechanical complexity. Therefore, the manufacture and use of these mechanical dry pipe valves has all but discontinued, and wherever possible, they are replaced in existing systems with simpler, more efficient, trouble-free, low or ordinary differential dry pipe valves.

Dry System Air Supply

Whether a dry system has its own air compressor or uses shop air, the air supply to the system must be regulated to maintain the required *system air pressure*. (See Figure 3.10.)

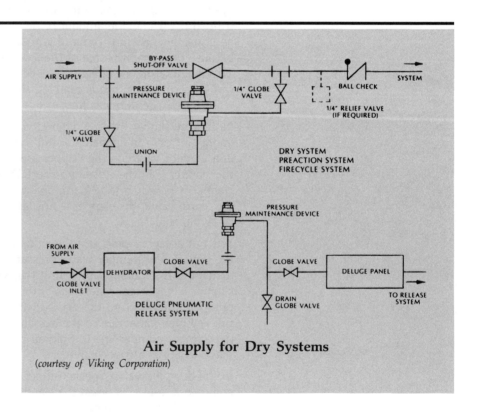

Air Supply for Dry Systems

(*courtesy of Viking Corporation*)

Figure 3.10

When a dry system has its own air compressor, this operation is performed by a *hi-low switch*. The hi-low switch is an air pressure operated electric switch connected to the dry system air supply. When the system air pressure drops below a predetermined point (still well above trip pressure), the pressure switch is activated and closes an electric contact which automatically starts the air compressor. This, in turn, supplies air to the system until the system pressure reaches the required pounds per square inch. The preset pressure again activates the hi-low pressure switch, and the electrical contact is broken, stopping the compressor.

The air supply for a dry system should be capable of restoring required air pressure in the system within 30 minutes in ordinary differential (5: or 6:1) dry pipe valve systems, and within 60 minutes in a low differential dry pipe valve system. The compressed air supply should be provided with at least one relief valve set to operate at a pressure 5 psi in excess of the maximum system air pressure.

Shop air (compressed air used in operating the plant) is not always within the acceptable pressure range of that required in a dry system, and due to facility use, often fluctuates. However, initial equipment and maintenance costs can be eliminated by using shop air in lieu of providing and installing a separate air compressor. The installation of a relatively inexpensive and maintenance-free device is an existing air maintenance system that can provide acceptable control of shop air. The air maintenance device is installed in an airline from the shop air system to the dry system, where it mechanically and automatically accepts the pressure the shop air delivers, and maintains the constant required air supply pressure to the dry system.

A note of caution: The pressure maintenance device will accept shop air pressure that is higher than the air pressure required in the dry pipe system piping. The device will automatically convert higher pressure to the lower constant pressure required for the dry pipe system, but cannot accept a shop air pressure lower than the dry system pressure and convert it to the required dry system pressure. In order to provide a continual acceptable pressure for fire protection, shop air must be maintained at a higher pressure than required for a dry pipe system. Another factor to monitor is during operation, ensure that the shop air used in the plant does not pull the pressure *below* the dry pipe system requirements.

Other maintenance considerations include plant shut down. When the plant is shut down for weekends, holidays, or at night, the shop air compressor cannot shut down; another condition where this could become a problem would be during a strike.

If shop air is not a reliable source of pressure, the next choices are an *air compressor* dedicated to supply air to the dry pipe system or a *small 1/4 horsepower maintenance compressor* to supplement the shop air.

The automatic air supply to the dry system cannot in any way interfere with the operation of the dry pipe valve by replenishing the air supply when a sprinkler head has operated and is releasing the air pressure. The air from the air supply is admitted to the system through an orifice which is so small that it cannot possibly admit enough air quickly enough to interfere with the tripping of the dry pipe valve.

Optional Devices

On large dry pipe systems, where temperature changes may produce periodic fluctuations in system air pressure, and where the possibility of the dry valve tripping could cause very serious problems, an *automatic air compressor* may be equipped with a *reservoir tank*. The air compressor maintains a constant air pressure in the tank and this tank air pressure floats on the dry system, which prevents the frequent cycling of the air compressor and increases the reliability of constant air pressure in the system. The tripping of a dry pipe valve due to low air pressure and the filling of the piping with water in an area subject to freezing is always a serious problem if not detected before the water freezes and cracks a fitting.

This problem quickly goes from serious to critical when this occurs with a dry pipe system protecting a *freezer*. The water will not only freeze very quickly, but if ice forms in the piping, and the freezer cannot be shut down until the piping thaws, it may even become necessary to disassemble the system, thaw out the piping, and reinstall the system, obviously a costly operation.

Inexpensive devices can be installed on dry pipe systems to protect it from dangerous air pressure fluctuations by initiating an alarm, such as a *low air trouble alarm*. Automatic air supplies are dependent on mechanical and electrical equipment which is naturally subject to failure. This failure potential is usually in direct proportion to maintenance of the equipment. If the automatic air supply to the system should fail and the system air pressure drops below a predetermined point, it would actuate the low air alarm pressure switch, electric contacts would close, and a trouble alarm would alert personnel to the need for immediate attention before the pressure drops to the trip point. These devices should supervise the dry system air pressure and activate a trouble alarm for either a predetermined low or high pressure setting. The *air supervisory switch* should be included when specifying a low air alarm device.

When the device is activated, and especially when it is activated by low air pressure, the trouble signal must be automatically transmitted to a constantly attended location. A low air trouble alarm means the water supply control valve for the dry system must be closed immediately until the problem is corrected to prevent serious damage to the system and/or costly maintenance.

Another optional device that may be required is a *dehydrator* on the air supply to the dry pipe system. The dehydrator is used to dry the pressurized air supply to the system to minimize the amount of moisture in the air supply, and thereby minimize the amount of condensation in the dry pipe system piping.

Pressurized Nitrogen Supply

Some dry pipe systems require the use of nitrogen as a more economical alternative than an air compressor due to a lack of available and/or adequate electric power supply. More often, nitrogen is considered when the air supply will contain considerable moisture or where the temperature differential between the air supply and the dry system is quite pronounced, and relatively warm, moist air would be introduced into extremely cold piping.

Nitrogen cylinders are connected to the dry system similar to the air pressure supply arrangement, with an air pressure maintenance device to maintain required system pressure. The use of nitrogen cylinders introduces an additional maintenance item since the supply is limited, and therefore, the capacity and pressure of the cylinders must be monitored regularly. The small dry system leaks that go unnoticed with air supplied by an air compressor or shop air can present a very critical situation where the pressurized supply is limited. A *reserve cylinder* of nitrogen must be readily available at all times to replace a cylinder that has fallen below required capacity.

Dry Valve Intermediate Chamber

The majority of modern dry pipe valves have an *intermediate chamber* which surrounds the water supply face of the clapper, but when the system is in the valve closed position, this intermediate chamber is sealed off from water and compressed air supplies. The function of the intermediate chamber is to speed the action of the dry pipe valve as follows: Most dry pipe valves are designed so that the initial drop in air pressure when a sprinkler head operates will allow the *clapper*, or an *auxiliary clapper*, to raise slightly due to the supply water pressure. Some supply water is then forced into the intermediate chamber under the main clapper. Now the air/water pressure differential is destroyed because the water pressure is acting on a much larger area—the water supply pipe size surface plus the intermediate chamber surface. Consequently, the air pressure is not capable of holding the clapper in the closed position.

> Modern dry pipe valves have various automatic methods of keeping the intermediate chamber dry at ambient atmospheric pressure. For example: Rockwood Firematic dry pipe valves offer a simple example of this device.

> When the dry valve clapper is in the closed position, it forces down a lever arm which rotates on a hinge pin and holds a drip valve raised above a drain piping opening. When the intermediate chamber fills with water and raises the clapper, it relieves this pressure on the lever arm, and the rush of water flowing through the drain opening forces the drip valve to close the opening. This action is similar to the ball drip drain described under the wet pipe system fire department connection, except that in this case, the drain is kept open by the action of the dry valve clapper on the lever arm.

Dry System Fire Department Pumper Connection

When the supply water pressure forces the main dry valve clapper to rotate on its hinge pin, it engages a latch mechanism which holds it in the open position. To reset this type of dry pipe valve, it is necessary to drain the system of water, unbolt and remove a face plate, manually release the latch, and free the clapper.

There is an interesting difference between the *alarm valve* and the *dry valve* which influences the design of a dry pipe system. In discussing the wet system, and specifically the fire department pumper connection to the system, the alarm valve served as a *check valve* to prevent the fire department from pumping back into the system supply line. Since the dry pipe valve clapper latches in the open position, the dry pipe valve *cannot* be used as a check valve, and a *standard check* valve must be installed in the main supply line.

Quick Opening Devices

Over the years, effort has been made to overcome the head actuation and water discharge time delay encountered with a dry pipe system. This has been accomplished, to some degree, by the installation of quick opening devices, to either increase the rate of discharge of air from the system piping, or to accelerate the opening of the dry valve.

If a dry pipe system contains in excess of 500 gallons, a quick opening device is required by the NFPA. These devices, depending on their method of quick opening the dry pipe valve, are *call exhausters* and *accelerators*. The operating principle of one such quick opening device is based on the use of two air chambers of equal pressure separated by a movable diaphragm. With this accelerator, one air chamber is designated as the inlet or lower chamber, and is connected to the dry system so that the air pressure in this chamber is equal to the dry system air pressure at all times.

The second chamber, called the *upper* or *pressure chamber*, is completely closed and sealed with the exception of a small, *restricted orifice* connecting the upper and lower chambers. This restricted orifice allows the pressure in the lower chamber to seep into the upper chamber so the pressure in the upper and lower chambers can become equal.

The two chambers are separated by a diaphragm, and when a sprinkler head actuates, there is a sudden drop in the dry system air pressure, and at the same time, in the lower chamber. The restricted orifice is much too small to allow sufficient air to escape from the upper chamber to compensate for this sudden drop in air pressure in the lower chamber, and the upper chamber, or original system air pressure, exerts an unequal force on the diaphragm forcing it downward. This movement of the diaphragm actuates various mechanical mechanisms and valves, the end result being the immediate tripping of the dry pipe valve.

Accelerators

The movement of the diaphragm in an accelerator opens an auxiliary valve which admits compressed air to the intermediate chamber of the dry pipe valve. This immediately upsets the water-to-air pressure differential of the dry pipe valve. The air pressure on the air side of the clapper can no longer hold the clapper closed against the water pressure, and the additional air pressure in the intermediate chamber and the system water pressure forces the clapper open.

In order to allow the air pressure in the intermediate chamber to exert its full force on the underside of the clapper, the intermediate chamber drain must close automatically when the compressed air is introduced. Latched clapper or mechanical-type dry pipe valves do not have an intermediate chamber and, therefore, cannot be tripped through the use of an accelerator.

Exhauster

The movement of the diaphragm in an exhauster opens an *auxiliary valve* on the device which dumps, or exhausts, the system air pressure to the atmosphere. The auxiliary valve on an exhauster is of such a large size that great quantities of system air are exhausted very rapidly and, as a result, the system air pressure drops quickly to the dry pipe valve trip point. Once the dry pipe valve has tripped and water starts to fill the system, water is transmitted to the exhauster. The action of the water, combined with the mechanical design of the exhauster in a wet condition, closes the auxiliary valve automatically, preventing discharge of water to the atmosphere through the exhauster valve.

Dry Pipe System Drainage

Because a dry system is used in areas subject to freezing, it is imperative to keep the piping empty of water. Piping is pitched to facilitate (and guarantee) gravity draining of any water in the piping back to the main drain at the dry pipe valve when the system is drained. The required piping pitch is at least 1/2" in every 10' for the branch lines, and at least 1/4" in every 10' for cross mains and feed mains. In refrigerated areas protected with a dry pipe system, it is good practice to increase this pitch as much as possible. An increase ensures that water does not accumulate from condensation, and that all the water drains out quickly after the system has tripped and is being drained for resetting.

Pipe Support

An additional point to consider in pitching the piping on a dry system is the possibility that the construction supporting the sprinkler piping may settle under the additional weight of the sprinkler system. This can occur in light construction, especially that which was not originally designed to accommodate a sprinkler installation.

Refrigerated Areas

In refrigerated areas, and areas where the possibility of settling exists, the piping should be pitched at least 3/4" to 1" in every 10' for branch lines, and at least 1/2" in 10' for feed and cross mains. As mentioned previously, the branch lines, cross mains, and feed mains in these areas should be pitched as much as construction and floor height will permit using the 3/4" to 1" in 10' for branch lines and 1/2" in 10' for cross mains and feed mains as minimum criteria.

The main drain on a dry system has identical size requirements to the wet pipe system, with the drain size based on riser size.

Special Conditions

Due to construction conditions such as low rooms, or decks, a dry system, like the wet system, may have low trapped sections of piping that will not drain back automatically by gravity to the main drain at the dry pipe valve. These trapped sections on a dry system are obviously of greater concern than those on a wet system; trapped water in the piping of a dry system can freeze, expand, crack the piping, cause a loss of air pressure, trip the dry valve, and cause serious and expensive water damage.

Another hazard is present when the dry system is shut down for repairs, which not only leaves the facility without fire protection, but also invites the possibility that a closed main control valve could be inadvertently left unopened, depriving the system of water supply for an unidentified period.

Maintenance Considerations

Maintenance for draining trapped areas on a dry system must be relatively easy and trouble-free. Maintenance personnel should readily drain these sections of piping after the system has tripped, or be able to determine if water from condensation has built up in the low piping.

Drainage Capacities

Where a trapped section of piping contains five gallons of water or less, the auxiliary drain for these areas may consist of a 1/2" valve with a brass plug or nipple and cap, to reduce leakage of system air through a defective drain valve. A trapped section on a wet system of similar capacity only requires a *plug* or *nipple and cap*. With the drain valve required for an auxiliary drain on a dry system, it is possible to carefully bleed off condensation water accumulation by barely cracking the valve without tripping it. This would be an impossible operation if a plug or cap had to be removed.

Drum-Drip Drain

Where the capacity of the low or trapped section of piping on a dry system contains in excess of five gallons, the auxiliary drain consists of a drum-drip arrangement. A trapped section of piping containing more than five gallons is large enough to allow condensation to collect, which could build up rather rapidly in the drain line. If not emptied frequently, this water could completely fill the drain line, freeze, and break a fitting. A drum-drip auxiliary drain is a collecting point for condensation water, an area which allows for a certain amount of expansion if the water should freeze. This provides for easy removal of a relatively large quantity of collected water without tripping the system.

Should the drum-drip fill, freeze, and crack the pipe or fitting in the condensation collecting nipple, the system would lose air and trip. Resetting the system back into service requires merely closing the upstream valve on the drum-drip, draining, and restoring the dry system to service, and then replacing the drum-drip assembly at the convenience of the maintenance department. No prolonged shutdown of the system is required.

A drum-drip auxiliary drain consists of two 1" valves, one on the system supply side of a 2" diameter by 12" long pipe, (or unit of equivalent volume), which is the condensation collector, and one on the discharge side of the condensation collector.

The valve on the supply side is kept open while the valve on the discharge side is closed and sealed with a brass plug or nipple and cap to prevent air from escaping should a valve leak. When condensation collects in the piping of a trapped section, it drains by gravity through to the 1" drain line, through the open 1" supply side valve, and collects in the condensation chamber. During regular maintenance, the supply side valve is closed, the discharge valve opened, and the plug or nipple and cap removed to drain out water that has accumulated.

Low Points

If several low points or trapped areas on a dry system are all located in the same general area, it will facilitate maintenance to install a tie-in drain. Connecting or tieing in all the low point drains through a common 1" pipe, and providing one auxiliary drum-drip drain at the termination of this tie-in drain, increases the reliability of all the low points being drained.

Dry System Design Features

This section will only deal with a few of the basic design considerations which are required when a dry system is installed.

Calculating the Area of Application

Depending on the design of the dry system, and not including unusually large supply pipe or long runs of supply pipe, the capacity of a dry system varies between 3/4 to 1-3/4 gallons per sprinkler. For an existing system, a good rule of thumb in estimating (not engineering) the capacity of a dry pipe system is to multiply the number of sprinklers by 1.25 gallons per head, and add the capacity of the bulk, or supply, piping if it exceeds 20' in length. This 1.25 gallons per head includes the line piping and the cross main piping when the dry pipe system layout is relatively simple.

When calculating the accurate capacity of a dry pipe system, the pipe of different sizes must be tabulated, totaled, multiplied by the number of gallons per foot capacity of each pipe size.

A system anticipating more than a 750-gallon capacity should be controlled by one dry pipe valve to compensate for the previously discussed time lag in discharging water on the fire once a sprinkler head is actuated. It is possible that this 750-gallon capacity will be exceeded in some dry pipe system designs, but not be enough to economically warrant an additional dry pipe valve. When this condition occurs, it becomes necessary to test the water delivery time following the tripping of the dry pipe valve, and this is done by opening *the inspector's test connection*. If the system is able to deliver water to the open inspector's connection within 60 seconds from the time the valve trips, and the inspector's test connection is located at a point most remote from the dry pipe valve, and at the highest elevation of the system above the dry pipe valve, the requirement for an additional dry pipe valve to carry the excess (over 750 gallon) capacity may be waived by the authority having jurisdiction.

Qualification test approval of an excess capacity dry pipe system cannot be based entirely on the 60-second water delivery. As with all fire protection systems, the evaluation of all the elements involved must be considered by a qualified individual. In this case, the 60-second water delivery is merely one facet of an acceptable dry pipe system. The occupancy being protected, construction material of the structure, floor heights, even the availability of the public fire department, their equipment, and their manpower, must be considered and evaluated.

Location of Dry Valves
The location of the dry valves is an important element in the design and function of the system. The dry valve or valves should be located as close as possible to the system to keep the amount of supply piping to a minimum and thereby reduce the total pressurized air capacity of the system.

The piping up to the clapper of the dry valve is filled with water, and in some dry valves there is a priming water seal on the dry side of the clapper in the dry valve. Naturally, this wet portion of the system must be kept from freezing. This can be done by locating the dry valve or valves in a warm area of the building, or locating them in a special dry pipe valve enclosure to protect the wet section of the dry system from freezing.

A dry pipe valve house may be constructed either inside or outside of the building being protected by the dry system. The valve house must be of insulated construction with walk-in refrigerator-type door and hardware so that a small capacity thermostatically controlled heater can maintain at least a 40°F temperature inside the valve house, regardless of the outside temperature. The heater may be electric, steam, or any other practical, reliable, and economical heat source.

To increase the reliability of maintaining adequate heat in the valve house or enclosure, a low temperature alarm device can be provided. Should the heat source fail to provide 40°F, the low temperature alarm will activate and transmit an alarm signal. As with the low air system on the dry pipe system, the low temperature trouble alarm signal should be transmitted to a constantly attended location for immediate attention.

Immediate attention is not quite as critical as it is with the receipt of a low air trouble alarm, because the insulated valve house or enclosure will retain sufficient heat to prevent a hard freeze for a short period of time, but attention to the problem must be given high priority. In addition to heat, the valve house must have adequate lighting to facilitate maintenance and inspection of the dry pipe valves, and be of a size that will provide easy access to all sides of the dry pipe valve and dry valve trim.

Water Columning

Dry system risers leaving the valve house or enclosure have been known to fill with water and freeze due to water columning. This is due to the accumulation of water in the riser resulting from leakage past the water seat of the dry pipe valve clapper or slow drainage of condensation back to the dry valve. These risers are usually large diameter pipes and when the water freezes there is enough capacity to absorb the expansion. However the ice in the pipe can render the system inoperative by preventing water passage in the pipe, or transporting the ice to small piping or sprinkler head orifices and blocking them.

The accumulation of water in the riser may not rise to a height in the riser past the roof or wall of the heated valve house and into the freezing area. However, due to the operational characteristics of the ordinary dry pipe valve, a relative minor accumulation of water in the riser can produce additional weight on the air side of the clapper and increase the trip point of the valve. When the system operates and the system air pressure is released, the valve will not trip as quickly because of the weight of water on the dry side of the clapper.

The dry pipe valves equipped with intermediate chambers overcome water columning. The automatic drain feature, which keeps the intermediate chamber open to the atmosphere, will not allow any water accumulation. This does not prevent slow drain back of condensation from the system from building up in the riser. In addition, there are dry pipe valves that do not have intermediate chambers, so each dry system design where water columning could present a problem should include an approved method of eliminating this water accumulation. This may be accomplished by installing a *high water level signaling device*, a *float-type automatic drain valve*, or an *automatic drain device*.

Low differential dry valves, because of their design function, are not subject to the weight of water on the clapper causing a time delay in operating. The possibility of the water accumulating and freezing in the riser makes it necessary to install a high water level signaling device or an automatic drain even though the dry valve is the low differential type.

One design feature which can make a dry pipe system more expensive than a wet system is the necessity for concealing the piping above a ceiling and installing the sprinkler heads below the ceiling in a pendent position, or when construction or head room mandates turning the heads in the pendent position on exposed piping. Only special dry pendent sprinklers may be used in the pendent position on a dry system. These special heads are described in detail in Chapter 2, "Sprinkler Heads," but basically, they are constructed to prevent water from collecting in the head or in the drop nipple to the pendent sprinkler head. When the system trips, water would naturally collect in the shaft of a head-turned pendent or in the drop nipple from concealed piping to the pendent sprinkler below the ceiling. This water would not drain off when the system was drained, and the dry pendent sprinkler prevents water from entering the pendent head shaft or drop nipple.

Pre-Action Systems

The dry pipe system was an adaptation of the basic wet pipe system; the pre-action system is a sophistication of the dry pipe system. The pre-action system provides more efficient fire protection than the dry pipe system and minimizes the possibility of water damage—both accidental and fire related—but these features introduce *additional cost*. (See Figure 3.11.)

When installing fire protection in a freezing environment, it is necessary to accept the added cost of the dry pipe system, both in equipment and labor of installation. Further, the additional cost of the pre-action system must be compared with the protection value for both fire and water damage that the pre-action system provides.

A pre-action system, except for the pre-action valve, resembles a dry pipe system, with the same piping arrangement, supports, sprinkler head coverage for the hazard protected, and closed head sprinkler heads.

However, the pre-action system has a second system: a detection system. The piping for a pre-action system is dry, but it cannot be referred to as a dry pipe system. Dry pipe systems contain piping with pressurized air or

Pre-Action System: A system employing automatic sprinklers attached to a piping system containing air that may or may not be under pressure, with a supplemental heat responsive system of generally more sensitive characteristics than the automatic sprinklers themselves, installed in the same areas as the sprinklers; actuation of the heat responsive system, as from a fire, opens a valve which permits water to flow into the sprinkler piping system and to be discharged from those sprinklers which were opened by heat from the fire.

Pre-Action Sprinkler Systems

Figure 3.11

nitrogen, and has a valve that is tripped by the loss of pressurized air or nitrogen in the piping. A basic pre-action system has a water supply that is held back by a pre-action valve. The fusing of a sprinkler head on the system will not activate the pre-action valve and admit water to the system. A piece of piping on the pre-action system can be removed and the pre-action valve will not trip, nor will water be discharged.

The only action that will trip a pre-action valve and admit water to the system piping is the activation of the detection system. This statement is not completely true because there are manual devices available to trip the pre-action valve, but the only automatic way to trip the pre-action valve is activation of the detection system.

The Pre-Action Valve

Water supply to the system is held back by the pre-action valve. When the detection system is activated: (1) A signal is transmitted to the pre-action valve, (2) the pre-action valve is tripped, and (3) water fills the piping. At this point, no water is discharged until the heat from the combustion fuses a sprinkler head.

A sprinkler head, even those temperature-rated at 135°F or 165°F, are not considered to be sensitive detectors. When the heat from combustion has risen sufficiently to fuse a sprinkler head, considerable damage can be inflicted on the occupancy before the sprinkler discharge suppresses the combustion and cools the area.

Detection System

By using a detection system that is much more sensitive than a sprinkler head to combustion in the very early stages, a fire alarm will be activated and personnel may be able to extinguish the combustion before sufficient heat develops to fuse a sprinkler head. The detection system that has been activated not only actuates the alarm system, but also trips the pre-action valve and admits water to the pre-action system. No water has been discharged at this point because none of the heads have fused, but the pre-action system has become a wet pipe system.

If the occupants cannot extinguish the combustion with hand extinguishers or hose, and the combustion continues to develop, sprinkler heads will fuse, water will be discharged from the open heads, and the combustion controlled or extinguished.

A pre-action system provides an early warning of a fire condition by the activation of the detection system in most cases long before there is sufficient heat to fuse a sprinkler head.

The detailed description of various detectors and detection systems will be found in Chapter 9, "Detection and Alarm Systems." In order to explain more fully the advantages of a pre-action system, two detection systems will be briefly described in this section covering pre-action systems.

An *ionization detection system* uses ionization detectors that will identify the products of combustion produced while the combustion is in the incipient stage—that is before there is visible smoke, noticeable heat, or odor.

A *rate-of-rise detection system* uses detectors that are activated when the temperature rises in excess of 15°F per minute, regardless of the ambient temperature of the area. When the detection system is rate-of-rise, the pneumatic tubing also maintains a small air pressure, and should a leak occur and the tubing pressure be released, two actions will take place simultaneously. First, a trouble alarm will be activated, and secondly, the pre-action valve will automatically be tripped admitting water to the piping.

Either of these sensitive detection systems will, in most cases, activate long before there is sufficient heat to fuse a sprinkler head or damage equipment. Should a fire in the area get out of control, the pre-action piping has been primed, full of water, and ready to discharge through the first sprinkler head that fuses. This early alarm warning will allow occupants alerted by the alarm to investigate and determine the source of the combustion, and take action to extinguish the combustion, shut off power to a smoldering wire or panel board, etc., before the combustion can develop enough heat to fuse a sprinkler head.

Protection of Equipment

When sprinklers must protect an area containing computer or electronic equipment, a pre-action system is recommended, even though water damage from a non-fire situation such as a broken sprinkler head or a leaking fitting is highly unlikely. Many objections are raised when an insurance carrier requires that a computer room be protected with automatic sprinklers. There are several good reasons for this requirement. For example:

> A computer room is situated in a building that is totally protected with automatic sprinklers, but there is no sprinkler protection in the computer room. Should a fire in the computer room get out of control, involve the entire room, and then break out into the sprinklered area of the building, it would, in all probability, overpower the sprinkler system, and involve the entire facility. When objections are raised to protecting the computer room or the electronics room with sprinklers, it can be stated that when a fire develops enough heat to fuse a sprinkler head, there is sufficient heat to damage or destroy all of the equipment in the room. When a fire reaches this condition, water damage is the least of all concerns.

Supervising the Piping System with Air

An additional safeguard against mechanical, non-fire, discharge of water with a pre-action system is to supervise the pre-action system piping with compressed air. In fact, the NFPA requires that pre-action system piping shall be supervised if the system serves more than 20 heads. In such cases, a small air pressure is maintained in the pre-action system piping, and this air pressure is monitored by a *low air pressure alarm device*. Should there be a small loss of air pressure through a defective fitting or pipe crack, the air pressure will drop to a predetermined point and activate the low air alarm. No water will fill the piping when the low air alarm is activated because, as previously explained, only the detection system can automatically trip the pre-action valve and admit water to the system.

The advantage of supervising the pre-action system piping is to prevent any leaks when the piping is filled with water. This could occur when the detection system has been activated and the piping has filled with water, but the combustion has been extinguished with an extinguisher before sufficient heat had developed to fuse a sprinkler head.

If the detection system is electrical, it is supervised to protect against a wiring break, short circuit, or ground. When one of these faults occurs, a trouble alarm is activated and the pre-action valve tripped, admitting water to the piping.

Supervision of the detection system is necessary because the detection system controls the sprinklers. When the detection system is rendered inoperative, the pre-action valve is automatically tripped, water fills the piping, and sprinkler protection is maintained until the fault can be corrected and the total pre-action system restored to operational status.

Use of Pre-action Systems

Since the piping of a pre-action system is dry, these systems can be used in areas subject to freezing. The same design criteria is used for a pre-action system in these areas: When heads are installed in the pendent position they must be dry pendent heads. When a pre-action system is used in areas not subject to freezing, regular pendent heads may be used, but only if their use is approved by the insurance interest.

Pre-action systems are not recommended for use in cold storage areas such as freezers or walk-in refrigerators because of the time element between detection, subsequent filling of the piping with water, personnel response, and possible extinguishment of the combustion before a head fuses.

Most early warning detection systems are not affected by low temperature. A rapid rise in temperature activating a rate-of-rise detection system, or the products of combustion with an ionization detection system, could fill the piping with water long before any action was taken to drain the pre-action sprinkler piping. This water in the piping with no movement, and the piping at the temperature of the freezer, would allow the water to freeze very quickly, which would present a difficult and expensive problem. The freezer temperature could not be lowered to below freezing long enough to allow the frozen water in the piping to turn back to water and be drained off. With the piping completely clogged with ice, the sprinkler system is useless. The only alternative is to remove the piping, thaw the ice, and reinstall the piping system.

During this expensive and time-consuming operation, the one fire protection advantage would be to keep the detection system in service.

The use of pre-action systems in areas subject to freezing is a matter of engineering judgment combined with a thorough familiarity with all the conditions surrounding the installation. The following is an actual case example where the human element was not considered.

> A pre-action system was installed to protect the coal conveyor of a steam power generating facility. A few years after installation, which was also a few years after operating instructions had been received and reviewed with the operating personnel (and several personnel changes had occurred) a bearing on a conveyor roller overheated and ignited the accumulation of lubricating grease. It was a small fire, but intense enough to activate the detection system, which opened the pre-action valve admitting water to the piping and transmitted an alarm to the control room. Personnel were dispatched to the area and the fire was quickly extinguished with a dry chemical extinguisher. Since the heat from the fire diffused by the extreme draft conditions often found in conveyor enclosures was not intensely concentrated enough to fuse a sprinkler head, no water was discharged. The sprinkler system looked no different after the

fire than it had looked before the fire, and no different than it had looked to the operating personnel every day over the years. No one realized, or was aware of the fact that, the system had tripped and should have been drained down and the valve reset. The result was obvious, the mechanical repair to the conveyor was completed in short order and the personnel returned to their normal duties. The piping froze overnight, and when the sun shone down on the metal roof of the conveyor enclosure the next day it thawed some of the piping and water began to gush from a fitting that had cracked under the pressure of the frozen water. Considerable water damage occurred before the leaks were discovered, and a considerable expense was involved in repairing the sprinkler piping. The human element must be a serious consideration in the judgment to use a pre-action system under circumstances where the reliance on adequate and informed operating personnel is a serious consideration.

Pre-Action System Design Features

As stated previously, a pre-action system has the same three essentials as the wet and dry pipe systems: alarm, water supply valve control, and fire department pumper connection. The difference is the pre-action valve, a deluge valve, or a *pre-action deluge valve*. The deluge valve is not activated by flow of water as is the alarm valve, or by a loss of pressurized air or nitrogen as is the dry pipe valve. Instead, deluge valves are activated or tripped by a signal from a detection system. A deluge valve is latched or held shut by water pressure and, regardless of what may affect the system piping or sprinkler heads, the valve does not trip and discharge water to the system until a detection system releases the valve to allow water flow.

Once a deluge valve is tripped and water flows, water can enter an alarm line and activate a pressure switch, close electrical contacts, and activate an alarm system. In addition, water can also be admitted to the alarm line to operate a water motor alarm. The pressure switch and water motor alarm are identical to those used on the wet and dry pipe systems.

A word of caution: The use of a *paddle-type flow switch* is *not* allowed on a pre-action system for the same reason this device could not be used on a dry pipe system. When the deluge valve opens, there is a rush of water into the system, and the velocity of this volume of water would seriously damage the paddle of the flow switch in the pipe. *Do not use flow switches on dry pipe or pre-action systems.*

Another important design feature of pre-action systems that have supervised piping is maintenance of a constant air pressure. A *rubber-seated check valve* is necessary to maintain air pressure in the system riser, and is installed on the system side of the pre-action valve. Rubber seated check valves are used so that the clapper of the check valve will form an air tight seal. With metal-to-metal check valve seats, there can be a tiny metal defect that would allow air pressure to escape, or a grain of sand or dirt can become lodged under the clapper and destroy an air seal.

The fire department pumper connection is connected into the pre-action system riser on the system side of the deluge valve between the deluge valve and the rubber-seated check valve. This connection is made at this point on the system so the fire department could pump into the system even if the deluge valve had not tripped and was in the closed position.

As with the dry pipe valve, the deluge valve does not act as a check valve to prevent the fire department from pumping back into the supply. It is necessary to have a check valve in the supply riser downstream of the deluge valve.

If the water supply control valve is to be installed in the pre-action riser, it is customarily installed below the deluge valve and above the water supply check valve.

There is always a check valve in the fire department pumper line to prevent water in the riser from discharging into the siamese connection when the deluge valve trips. This check in the pumper line is required regardless of the type of system it serves. A small, usually 1/2" *automatic ball check valve* is installed on the siamese side of the check valve at the low point to automatically drain off any condensation water or leakage past the check valve to prevent freezing in the siamese.

Types of Pre-Action Systems

Basic Pre-Action System

What has been described is the basic pre-action system, and what one manufacturer refers to as a *single interlock pre-action system*. A detection system which can be electric, pneumatic, or hydraulic, automatically activates a pre-action valve and water is admitted to the system piping, but no water is discharged until a sprinkler head is activated. The fusing of a sprinkler head independent of the detection system will not initiate the filling of the piping with water, and therefore no water will be discharged from the activated head until the detection system is activated and trips the pre-action valve.

System piping may be supervised with pressurized air or nitrogen to detect leaks, broken piping, or a defective or fused sprinkler head before the pre-action valve is tripped and water enters the system. A low air trouble alarm will indicate the loss of the supervision pressurized air or nitrogen. Only activation of the detection system, manual operation of the emergency release, or malfunction of the detection system will admit water to the system piping.

Double Protection Pre-Action System

The double protection pre-action system is used in areas where maximum protection against accidental filling of the piping with water is required—such as refrigerated storage areas, sensitive electronic equipment areas, or computer rooms. One manufacturer refers to this type of pre-action system as a *double interlock pre-action system*. This system identification is mentioned here because it is frequently used to identify system design regardless of the manufacturer.

With the double protection dry-action system, both the detection system and a sprinkler head must be activated before the water will enter the system piping. Activation of either the detection system or a sprinkler head fusing will only activate an alarm.

The piping for this system contains pressurized air or nitrogen similar to the dry pipe system. Its design utilizes a deluge valve (pre-action valve) which can only be opened when pressure is reduced in the sprinkler piping and the detection system operates.

This system must only be used with the approval of the insurance interests having jurisdiction, and designed by experienced fire protection engineers because it greatly increases the non-fire water damage reliability and the potential for delay in applying fire suppression water discharge.

Combined Dry-Pipe and Pre-Action System

The combined dry-pipe and pre-action system can be activated by operation of the detection system or the fusing of a sprinkler head. One manufacturer's terminology for this system is the *regular non-interlock pre-action system*. One major design feature of this system is the use of a dry pipe valve in lieu of the deluge valve, but it differs from the basic dry pipe system because of two factors: Both the release of pressurized air or nitrogen in the piping, and the activation of the detection system will trip the dry pipe valve.

One major advantage of this system is that, should the detection system fail to operate or malfunction, the system can operate as a dry pipe system. By increasing the fire protection reliability, this system decreases the non-fire water damage potential. Should the sprinkler piping or a sprinkler head by damaged, the pressurized air or nitrogen will be released, the dry valve will trip, piping will fill with water, and the water will be discharged.

The advantage of this system over a basic dry pipe system is the detection system sensitivity which will provide an early warning alarm. Activation of the detection system will trip the dry valve and fill the piping with water, but water will not discharge until a sprinkler head fuses.

As previously described, the basic pre-action system valve must receive a signal from the detection system to allow water to flow in the piping, but with this system either the detection system or the fusing of a sprinkler head will activate the system.

Quoting from the National Fire Protection Association Standard for Installation of Sprinkler System:

> "Combined automatic dry-pipe and pre-action systems shall be so constructed that failure of the dry-pipe system of automatic sprinklers shall not prevent the fire detection system from properly functioning as an automatic fire alarm system."

Deluge Systems

Although sprinkler systems have an admirable performance record of property protection, some hazards require special protection in addition to the building protection provided by the sprinklers. This is evidenced by the relatively small number of serious fire losses that have occurred where combustion heat, intensity, and spread overpowered the sprinkler system. Areas where easily ignitable and fast burning material exist require special hazard systems.

The first special hazard system to be considered is the deluge system. (See Figure 3.12.) The wet and dry pipe systems previously described have closed sprinkler heads, heads with fusible links, or fusible elements that only discharge when the fusible link or element is heated to a predetermined temperature.

A deluge system consists of open sprinkler heads. The sprinkler heads on a deluge system are similar to the heads on a wet or dry pipe system, but they do not have any fusible links or elements. When the system is activated, water is discharged from every head on the system, simultaneously creating a deluge of water.

The piping on a deluge system is obviously empty because all the sprinkler heads are open. The deluge valve prevents the water from entering the system piping until a signal from the detection system trips the deluge valve. A deluge system valve is opened to admit water to the system by the action of a separate detection system. This detection system may be electric, pneumatic, or hydraulic.

The deluge system is designed to wet down the entire area protected by the system. One typical example is the protection of a cooling tower. For example:

> By virtue of its design, there are numerous wind currents surging through a cooling tower, and this is especially true when it is not in operation and dry. With a wet or dry pipe sprinkler system, the heat from a fire in the northwest corner could easily be deflected by wind currents and open closed heads in the southeast corner. Obviously, the discharge from these fused heads would have no suppression effect on the fire origin. With the deluge system valve activated by the detection system in the cooling tower, no matter which detector is activated (in this example, the heat would activate the detector in the southeast corner), the deluge valve will be tripped and water discharged over the entire area, extinguishing the combustion in the northwest corner.

Deluge System: A system employing open sprinklers attached to a piping system connected to a water supply through a valve which is opened by the operation of a heat responsive system installed in the same areas as the sprinklers. When this valve opens, water flows into the piping system and discharges from all sprinklers attached thereto.

All areas are assumed to be open.

Deluge Sprinkler Systems

Figure 3.12

Another advantage of a deluge system is its cooling capacity. The deluge of water not only controls and extinguishes the combustion, but also cools all equipment and structural members within the area of the system discharge. This prevents damage of equipment by radiant heat, or structural steel deformation by combustion heat possibly resulting in severe building structural damage.

The first deluge valve to consider is a quick opening valve having only one moving part, the *clapper*, which is kept closed by water pressure. This valve can be activated by a manual release or a hydraulic, pneumatic, or electric detection system. The deluge valve (produced by Viking) has three chambers: the top chamber which is under water pressure, the discharge outlet which is normally dry, and the inlet chamber or incoming water supply. The chambers are separated by the valve clapper. The water pressure on the top of the valve clapper, or top chamber water pressure, is the same as the inlet chamber or water supply pressure. This is accomplished by a small *priming line* which takes supply from the water supply and bypasses the clapper, discharging into the upper chamber. This priming line is referred to as the *bypass line*.

The priming line has a control valve, a restricted orifice to limit the flow of water through the line when the system is activated, and a strainer to prevent the passage of small particles of sand or dirt which could clog the restricted orifice.

The top of the valve clapper, which forms the bottom of the top chamber, is twice as large in area as the bottom of the clapper where it covers the top of the water supply inlet chamber—a 2:1 area differential design.

When the valve is in the closed position, the supply water pressure is equal on both sides of the valve clapper. On the inlet side, this water pressure is acting on approximately 28 square inches for a 6" water supply to the valve, while the same water pressure is acting on approximately 56 square inches on the top of the clapper. With a supply pressure of 75 psi (pounds per square inch), approximately 2100 pounds of supply water pressure pushing up on the bottom of the valve clapper is opposed by approximately 4200 pounds pushing down on the top of the valve clapper. (These figures are used for example only, and are therefore approximations.)

When the detection system is activated, the water pressure on the top of the clapper is relieved through the release line, and the clapper is lifted by the supply water pressure underneath. Water fills the piping and is discharged from all the sprinkler heads on the system. The priming, or bypass, line restricted orifice cannot make up water pressure fast enough to overcome the released pressure on the top of the clapper when the detection system has opened the release line and relieved the pressure on the top of the clapper.

Most other deluge valves have a clapper that is *latched* in the *closed position*. One such valve has a diaphragm with equal pressure on both sides. Activation of the detection system relieves the pressure on one side of the diaphragm and, when the diaphragm moves, it operates a shaft and lever mechanism which unlatches the clapper.

Another manufacturer provides a deluge valve that has a weight held in a fixed position. When the detection system is activated, the weight is released and drops, opening the clapper latch. The water pressure rotates the clapper to the open position.

The *water supply* to the deluge system is controlled by a valve in the water supply line to the system. As with the dry pipe system, once the system has tripped and the clapper has rotated to the open position, any water pumped through the fire department connection will pass back through the deluge valve into the supply, if the fire department is pumping water in at a higher pressure than the system supply pressure. Therefore, it is necessary to install a check valve in the water supply to the deluge system below the deluge valve to prevent the fire department from pumping back into the water supply. The fire department pumper line will be provided with a check valve to prevent water from leaking out of the siamese connection, or from discharging out of an uncapped siamese, when the deluge system is activated and water fills the system.

When the deluge valve is tripped and the system piping fills with water, the small alarm line is also filled with the pressurized water. This alarm line can supply a pressure switch which will make an electrical contact when the pressurized water enters the device, and can also operate a water motor alarm.

Since the clapper on deluge valves is held closed by a mechanical latch or water pressure, a retard chamber is not necessary. It is used in other systems to prevent supply water pressure surges from activating the alarm devices. Another device that cannot be used is the *paddle-type water flow switch*. Because the water rushes into the system piping when the deluge valve is tripped, the paddle in the pipe would either be ripped off or severely damaged by the velocity of a sudden rush of water. (This condition also exists on the dry pipe system.) Paddle-type water flow switches can only be used on wet pipe systems where the piping is constantly filled with water. When the system operates, the water surrounding the paddle in the piping merely deflects with the movement of the water.

Standpipe

Standpipe systems are piping systems that provide fire hose stations for manual application of water to fires in buildings. Standpipes are classified in three classes.

Class I Standpipe

This category of standpipe supplies 2-1/2″ hose outlets at each floor level, usually located in stairtowers, for use by firefighters or those (thoroughly) trained in handling heavy 2-1/2″ hose fire streams. These 2-1/2″ outlet hose valves furnish the firefighters with a hydrant at each floor level, and an adequate water supply to provide an effective hose stream for use during the advanced stages of a fire. A live hose outlet at each floor level installed in a fire-resistive stairtower to protect the fireman ascending to the fire floor allows the fireman to carry dry hose up the stairs in lieu of dragging a connect hose up the stairs. The fireman can control the flow of water from the floor below the fire floor to reduce the possibility of fire involvement in the stairtower at the point of hose connection when the stairtower door is opened on the fire floor.

Water Supply

Class I standpipes must have an adequate water supply to furnish sufficient volume and pressure for 2-1/2″ hose streams. The minimum water supply for Class I standpipes must provide a minimum of 500 gpm if only one standpipe is required, and 250 gpm for each additional standpipe. For example, if a building requires four standpipes, the water supply must be capable of providing 1250 gpm. The total supply demand shall not exceed 2500 gpm, and based on the 500 gpm for the first standpipe and 250 gpm for each succeeding standpipe, a maximum of nine standpipes.

Pressure Requirements

The pressure requirement for Class I, Class II, and Class III standpipes is a water supply that will provide 65 psi residual pressure with the required gpm flowing at the highest outlet on the standpipe. Residual pressure is the pressure reading on the gauge with the required gpm flowing. (Residual pressure will be discussed in Chapter 5, "Water Supplies."

Class II Standpipe

A Class II standpipe system is one designed to be used by the occupants of the building until the arrival of the fire department. It is sometimes referred to as a *small hose standpipe* because it supplies 1-1/2" hose stations. Note hose *stations*, not hose *valves*, are supplied by a Class II standpipe system. It is extremely dangerous to have 2-1/2" hose station with hose connected to the 2-1/2" hose valves on a Class I standpipe. If untrained individuals attempt to use the hose to fight a fire, they could be seriously injured from the excessive thrust produced by the flow of water under pressure. Handling 2-1/2" hose streams is a job for trained firefighters; not a one man job.

Some city building codes require 2-1/2" hose at each outlet to eliminate the need for firefighters to carry the hose up the stairs. Although a good concept from the firefighters' standpoint, there should be some latitude for judgment because there is always a good likelihood that an occupant could attempt to fight a fire with the 2-1/2" hose. The possibility of serious personal injury and unnecessary heavy water damage may outweigh the benefit to the firefighters.

There is another important reason for not furnishing 2-1/2" hose. The 2-1/2" hose outlets on the standpipe are usually (and should be) located in a stairtower that is constructed to provide safe exit for occupants and a safe access to the fire floor for fire fighters. If an occupant connects the 2-1/2" hose to the hose valve and drags the hose through the stairtower door to attempt to fight a fire in the building, the hose will prop open the stairtower door allowing smoke and heat to enter the stairtower which could make the stairtower impassable for occupants attempting to exit the building.

Even a small amount of smoke and heat in an exit stairtower can cause panic, which can contribute to death and injury as easily as fire. People coming down the stairs in the stairtower from the upper floors may trip over the extended hose in their haste to escape.

One more consideration against placing a supply of 2-1/2" hose by the 2-1/2" hose valve is the likelihood of an untrained occupant dragging the hose to the fire area in the building and, when he opens the nozzle, the hose is whipped from his hands. In the confusion of a live hose crashing about, the novice firefighter will run for the exit. By the time the fire department arrives, they are faced with a fully involved fire on the floor of fire origin, a partially opened stairtower door on the fire floor making access to the hose valve at that floor level impossible, and a 2-1/2" hose wasting water and reducing water supply volume and pressure needed to combat the fire.

Water Supply

The water supply for Class II standpipe service must be capable of supplying 100 gpm at the highest outlet on the standpipe with a residual pressure of 65 psi with the 100 gpm flowing.

A Class II standpipe supplies the 1-1/2" hose rack or hose rack in a cabinet, that are commonly found in commercial, industrial, and institutional facilities. These hose stations provide a means of controlling a fire in the early stage when used by the occupants of the building.

Pressure Requirements

One hundred psi is the maximum pressure allowed at any hose valve outlet on a standpipe system. This maximum pressure requirement is especially important to safeguard the untrained occupant using a 1-1/2" hose stream for the first time. Even with a 1-1/2" hose, high pressure discharge in inexperienced hands can make control extremely difficult and dangerous.

Standpipes that have a water supply that provides a standpipe pressure in excess of 100 psi often occurs when a standpipe serves a multi-story building or is supplied from a fire pump. To ensure a pressure of 65 psi at the top-most outlet of 100' high building means that a pressure of at least 143 psi must be provided at the first floor. Disregarding friction loss to transport water to the 100' hose outlet and merely adding the static pressure of 0.434 psi per foot which must be overcome to lift the water up to the 100' elevation. Pressure and flow will be fully explained in Chapter 8, "Hydraulic Calculations."

To reduce the various pressures at the hose outlets on different floors to 100 psi or below, special hose valve equipment must be installed. The simplest method is to install a pressure-reducing disc in the hose valve outlet. This disc consists of a brass plate with an orifice of a predetermined size drilled in the center. The pressure reducing disc is quite efficient if the standpipe pressure is relatively constant because the orifice size can be determined by a calculation of the standpipe pressure at the specific hose outlet. For example:

> To reduce the pressure at the nozzle on a 75' length of 1-1/2" hose to 80 psi when the standpipe pressure at the hose valve is 150 psi requires a 3/4" orifice. At 200 psi, it needs a 5/8" orifice, and at 125 psi, a 7/8" orifice is used.

Current state-of-the-art practice is to use a hose valve that is not factory set for a specific pressure reduction, but that automatically adjusts the inlet pressure to a safe outlet hose pressure.

The Class II hose stations should not be located in an exit stairtower for many of the same reasons the 2-1/2" hose should not be located there: The obstruction the hose would present to people using the stair to exit the building, and the propped-open stairtower door on the fire floor.

The hose valve referred to can be either an *angle valve* or *straightway valve* connected to piping from the standpipe riser. This valve can be opened before the hose is removed from the hose rack without emitting water from the nozzle. The folds in the hose prevent water from flowing until the last fold comes off the rack. Water will then flow from the nozzle unless it has a device to control the flow. Some nozzles not only have an on-off control, but also a means of changing the flow from a *straight stream* to a *water-fog*, and some of these combination nozzles have a method of adjusting the water-fog pattern. The selection of a nozzle for 1-1/2" hose is extremely important and must be based on a thorough evaluation of the hazard and user.

Improper application of a straight stream hose discharge has the potential to spread the combustion by scattering the burning material. For example, when directing the stream from a garden hose into a pocket of mud, the mud is scattered in all directions. The same situation will occur when an 1-1/2" straight stream is directed onto burning combustibles like paper, trash, etc. For example:

> An accidental flammable liquid spill that was ignited should never be fought with a straight stream water discharge. Like the mud puddle, the burning liquid will be scattered over a wide area.

> With this fuel spill fire, a combination nozzle, and an individual with some training in the use of small hose firefighting, the fuel fire can be controlled with the water spray nozzle adjustment. In addition, the straight stream nozzle adjustment can provide a discharge onto material and equipment that is exposed to the radiant heat of the combustion to maintain a temperature that will prevent heat damage and/or prevent flammable liquid drums and tanks from exploding.

Some nozzles discharge only a very fine spray pattern of tiny water droplets. This water pattern is like an extremely heavy duty fog, thus one trade name of *water-fog*. The spray pattern absorbs so much combustion heat that a person can advance close to even an intense fire without being overcome by the heat. This fog pattern will not scatter the burning material, and the water droplets in changing to steam absorb sufficient heat to control and extinguish the combustion.

Another important consideration is the presence of live electrical equipment. A solid straight stream discharging onto electrical equipment, to especially high voltage equipment, can act like a wire and conduct the current back to the nozzle and the user.

Unlike the straight stream, discharge of the spray pattern on electrical equipment does not have the potential to conduct an electrical current. However, the voltage of the electric equipment and the closeness of the nozzle to the equipment could produce an electrical shock to the hose operator.

Small hose nozzle selection deserves a thorough evaluation and engineering judgment. Small hose nozzles should not be a specification shelf-item selection.

The hose rack on which the hose is hung is usually a semi-automatic type. The hose is hung on pins, and it is released as the nozzle is pulled away from the rack.

1-1/2" Hose Stations

Small fires can be extinguished with a small hose before they grow into a major conflagration, or can be controlled until the fire department arrives. Hose stations with 1-1/2" hose for occupant use may be a curse or a blessing. Time is the greatest ally of fire, and unless there is a water flow alarm on a Class II standpipe that will automatically alert the fire department and activate local alarms, there is always the possibility that an occupant will not call the fire department right away. In an attempt to fight a fire, they will not call until realizing it is beyond their ability to so do, and then summon the fire department and alert other occupants. Meanwhile, the fire has ample time to develop before the firefighters arrive or the alarm is sounded.

These advantages and disadvantages must be given serious consideration when a standpipe system is being evaluated.

Class III Standpipe

The third standpipe system is called a *combination standpipe* and is classified as a Class III standpipe. The Class III standpipe system is a combination of a Class I and Class II standpipe. The Class III standpipe supplies both 2-1/2" hose outlets at each floor level in the stairtower, and Class II 1-1/2" hose stations located throughout the building outside of the stairtower.

On Class I and Class III standpipes it is sometimes required, and always a good idea, to install a 2-1/2" by 1-1/2" removable reducing coupling in the outlet of the 2-1/2" fire department valve. With this coupling, the fire department can connect their 1-1/2" hose or remove the coupling and connect their 2-1/2" hose. The coupling is usually a pin lug type with two pins that fit the hole on a spanner wrench for quick removal of the coupling.

Standpipe Water Supply

The water supply for standpipes can be furnished by a connection to a city water main, through a fire pump to boost the pressure, gravity tank, or pressure tank. If the supply is from a limited source—such as a tank in lieu of an unlimited supply from a city water main—it must be of adequate capacity to furnish the total demand for at least 30 minutes.

Types of Standpipes

There are two types of standpipe systems: a wet standpipe and a dry standpipe. (See Figure 3.13.) Like wet sprinkler systems, a wet standpipe pipe is constantly full of water under pressure and is used when the area is heated to prevent freezing.

There are two types of dry standpipe systems. One type is dry because it is not connected to any permanent water supply. A standpipe with fire department 2-1/2" hose valves at each floor level provides firefighters with an outlet when the fire department pumper truck pumps water into the standpipe riser(s) through a pumper connection similar to the siamese connection on a sprinkler system.

The total demand for a standpipe being used by the fire department may be more than a standard 2-1/2" × 2-1/2" × 4" siamese connection can provide. Depending on the size and number of standpipes, the fire department pumper connection on either a wet or dry standpipe may have four 2-1/2" outlets manifolded into a 6" supply pipe (refer back to Figure 3.4).

The dry standpipe without a permanent water supply is used in unheated areas, areas where the water supply is questionable, and buildings under construction where the pipe is exposed and could freeze or where the water supply is not completed during construction. Dry standpipes used during construction may remain dry or be converted to a standard wet standpipe when the construction is enclosed, heated, and a water supply connection complete.

When a dry, no water supply, standpipe is installed, and rises several floors above the pumper connection, a *check valve* should be installed in the main supply line on the standpipe system side of the pumper connection. With the riser(s) full of water under a static head, or elevation pressure, the small clappers in the siamese connection cannot be expected to retain the water under pressure. When hoses are removed from the siamese, a backflow can result. The check valve can hold water in the riser(s) until it can be safely and conveniently drained off through a drain valve installed on the standpipe system side of the check valve.

An important feature which must be incorporated into wet standpipe design for all three standpipe classes is a *main control valve* at the base of each standpipe riser, on the water supply side of all hose outlets on each riser. This valve is used to control the water supply to each riser in case of a break in the standpipe or if a hose that is discharging has been abandoned because of some unexpected firefighting condition, (as in the example of the occupant using a 2-1/2" hose). A control valve on each riser gives the fire department control of each standpipe riser if, for any reason, water volume and pressure is being wasted. Another reason for this individual riser control valve is that should a riser be down for repairs, additions, or alterations, only that one riser will be shut off and the others will remain in service and ready for immediate use.

The second type of dry standpipe is also used in unheated buildings or stairtowers, and is sometimes referred to as a *remote control fire hose standpipe*. Except under unusual circumstances, the dry standpipe with no permanent water supply should only be a Class I standpipe.

The remote control dry standpipe system design is similar to the wet Class I, Class II, or Class III standpipe, but the piping is empty. The water is held back by a deluge valve installed in a warm area of the building or a heated valve house.

Roof

Roof connections with hose gate valves (for combustible roof)

Hose connections on each floor (size based on class of service)

Check valve

Siamese inlet connections (for fire department use)

Dry Standpipe Riser

Figure 3.13

A remote control fire hose station, like a break glass manual fire alarm station, is installed immediately adjacent to each hose station. Each of these remote control stations is connected to the release mechanism of the deluge valve, and water fills the system when the manual station is activated.

Some remote control systems electrically activate the deluge valve release mechanism, but to eliminate the possibility of power failure making the standpipe system inoperative, several manufacturers have developed pneumatic systems. Operation of a remote manual station releases compressed air which, in turn, trips the deluge valve and water fills the standpipe system.

One dry standpipe system consists of remote control stations that function as follows: When the small glass window on the face of the remote manual box is broken with a small hammer (which is attached to the unit by a chain), a spring is released which operates a bellows sending an air pressure impulse to the release mechanism of the deluge valve.

If *air tight hose valves* are used on a dry standpipe system, it is possible to use a regular dry pipe valve to control the water supply with the piping filled with air under pressure. When a hose valve is opened, compressed air is released, and the dry pipe valve releases water to the system.

The remote station deluge valve system and the dry pipe valve system standpipes introduce an element that is not encountered with a wet pipe standpipe. Once the dry system is activated by operation of a manual remote station or opening a hose valve on the dry pipe valve system, there is a *delay* in water reaching the nozzle. This "charging period" can produce some unusual hose action when it is suddenly pressurized with water. A test conducted on a dry standpipe using remote stations (the spring and bellows) indicated that 11 seconds after activating the remote station 333' from a 6" deluge valve in the basement of the building, water was discharging from a 2-1/2" hose line. It is obvious that water rushing through piping and into hose at a velocity of 30' per second or more can create hose and nozzle reaction that could be dangerous, especially in inexperienced hands. If an individual is knowledgeable of the operation of the system, the remote station can be activated leaving the hose valve closed until the system fills with water and then opening the hose valve and removing the hose. Once the water reaches the closed hose valve, the system is a wet pipe standpipe system.

Hose Station Distribution

The distribution of hose stations is a matter of engineering judgment. Hose stations should be located so that each section of the protected area, and all portions of each floor, are within 30' of a nozzle attached to 100' of hose. Hose for use by occupants longer than 100' becomes unwieldy, could lead to twisted or kinked hose limiting the flow, given the fact that all the hose must be pulled off the rack before water will flow. This location of 1-1/2" hose stations is a minimum requirement, but the number of standpipes, location, and distribution of hose stations depends on occupancy, construction of the building, accessibility of hose outlets, and number of types of hazards. (See Figure 3.14.)

The standpipe risers with 2-1/2" hose outlets should be located where they will be protected from mechanical and fire damage and exposure, and readily accessible in a fire condition with the center line of the hose outlet approximately 2'-6" above the floor.

Each occupant-use hose station should consist of a 1-1/2" hose valve supplied from the riser. An automatic drip drain must be installed at the hose valve when linen hose is used to prevent any leakage past the hose valve from entering the hose. The hose nozzle may be a straightway pattern with no shut off valve and a 3/8" or 1/2" orifice. When the hose is pulled off the rack with the hose valve, open water will discharge.

The hose should be rubber or Neoprene lined, as lined hose greatly reduces the friction loss, with an outside woven jacket. Unlined linen hose has been used quite extensively because it folds easier and fits into a smaller space, and initial cost is less than lined hose. Unlined hose will mildew and rot if left damp or if stored in a damp atmosphere, and unless dried in a professional manner it is, in reality, a one-shot use hose.

Firefighters usually do not know if the hose stations are stocked with unlined hose that may crumble when pulled off the rack, or even lined hose that may be old and in an unreliable condition. In most cases they will use their own 1-1/2" hose.

If unlined linen hose is stored in a dry, heated atmosphere, it should last indefinitely, and for emergency use it is the most economical hose. On a construction project to serve the hose outlets of a standpipe that is installed as the building construction progresses, and with no intent to use the unlined linen hose in the finished building hose cabinets, unlined linen hose will serve an excellent, economical, purpose. This hose is usually constructed of 100% linen flax yarn to meet Underwriter's Laboratory (UL) listing and Factory Mutual (FM) approval requirements, and it can be given a mildew treatment to retard deterioration.

Most lined hose has a cover of cotton wrap with a Dacron or polyester filler, either single or double jacket. The lightweight polyester single jacket, neoprene-lined (tube) can be used on pin racks, and occupies only a little more room on the rack than unlined linen hose. Some hose

Hose Station

Figure 3.14

construction like single jacket and double jacket consisting of cotton/polyester with a neoprene tube lining cannot be folded to go on a pin rack and must be stored on a hose reel or hump rack. A hump rack is a rack where the hose is piled in horizontally, one fold on top of the next.

For many years, the size of the standpipe was governed by the height of the standpipe as follows:

A. Standpipe not exceeding 75' in height or 100' in length, 4" in size.
B. Standpipe in excess of 75' or 100' length, 6" in size.
C. Standpipe height limit 275' and when a building is in excess of 275', the standpipe system will be zoned. (Standpipe zoning in highrise buildings is explained at the end of this chapter.)

Note that the sizing requirements use the phrase, "in height or length." A piping system supplying hose stations throughout a one-story building is also referred to as a standpipe system.

With the advent of extensive use of hydraulic calculations for sprinkler systems in lieu of the pipe schedule sizing, it is recommended that standpipe systems be afforded the same engineering status. By hydraulically calculating a standpipe system, the pipe sizes can usually be greatly reduced, but of more importance than this economic advantage is the fact that calculations will determine that the flow and pressure at each hose outlet will be adequate to meet requirements.

Hose racks in buildings where appearance is a factor, are usually stored in cabinets that are completely recessed in the wall with the door flush with the wall, or semi-recessed with 1" to 3" extending beyond the wall surface. Surface mounted cabinets are not recessed into the wall and are mounted on the surface of the wall or in a column.

When designing a standpipe system and locating hose cabinets, it may be necessary to recess a cabinet in a wall that is rated to resist fire spread. Fire walls are constructed to prevent fire from spreading from one area to another for a specified time. These rated walls enclose stairtowers to protect people from the heat of a fire and smoke while they are exiting the building. Recessing a cabinet in a fire wall in effect creates a weak point in the fire wall. Should the thin metal cabinet be distorted by the heat of the fire, the cabinet can drop out of the fire wall and allow smoke and heat to pass through, involving personnel and material on the protected side of the fire wall.

If it becomes necessary to locate a recessed cabinet in a fire wall, the wall can be extended around the back, top, and bottom of the cabinet. If the cabinet should be lost, the integrity of the fire wall would not be destroyed.

The center line of a 1-1/2" hose valve should be located 5'-6" to 6' above the finished floor, for ease of operation. Unless local conditions dictate a different arrangement, the hose valve is usually an angle valve with a nipple, called a *rack nipple*, which screws into the valve outlet and supports the hose rack. The hose coupling connects to the rack nipple. The hose rack swings on the rack nipple a full 90°, and this operating distance, combined with the cabinet door swing of 170° (to give full and unobstructed access to the hose rack in the cabinet), must be considered in designing the location of racks and cabinets.

Some hose cabinets are sized to accommodate both the 1-1/2" hose and a separate 2-1/2" hose valve. Also available are cabinets for individual 2-1/2" hose valves.

Roof Hose Stations

Some building codes require hose outlets on the roof of a building to provide firefighters with a hose outlet to fight roof fires ignited by embers from an exposed fire, a fire in an elevator, or mechanical penthouse, or to direct water to a fire in an adjoining building.

These roof outlets—known as roof manifolds—consist of two or three 2-1/2" hose valves with caps over the openings to prevent foreign matter from falling into the valve. The supply for these roof manifolds is a 4", 6", and 8" extension above the roof of the standpipe riser. Since wet standpipes and the hose manifolds are exposed to the weather and subject to freezing, an indicating control valve is installed at hand height in a heated area at the floor level below the roof and opened when the roof manifold is used. An automatic drain is installed on the roof monitor side of the control valve to make sure the pipe and roof monitor remain dry when the control valve is used.

The roof manifolds are 2-1/2" hose valves because they are hydrants on the roof for use by the firefighters. If there is a possibility of fires at the roof elevation that could be controlled or extinguished by building occupants or a company fire brigade, roof hose storage cabinets are available for 1-1/2" hose. These roof hose stations are supplied from piping on the ceiling of the floor below, but instead of controlling the supply to the roof hose station from the floor below the roof, the valve is installed in the piping on the ceiling below the roof. The handle of the valve extends through the roof and into the roof hose cabinet. The drain valve is installed like the handle, also extended into the hose cabinet.

The cabinets contain 100' of 1-1/2" lined, hose and nozzle, with the hose folded horizontally under a waterproof cover that unlatches and lifts up on hinges.

Standpipe Zoning

The maximum height for standpipe systems is 275', and when a standpipe must exceed this height, the system must be zoned. A zone signifies the upper and lower levels of a building served by standpipes. When zoning a standpipe, 275' is maximum height, but the zones may be less than 275'.

An example would be a building 355' in height. The lower level, or zone, could be 275', with the upper level of 80', but a more equal zoning would be a lower zone of 200' and an upper zone of 155'. If the two zones had exceeded 550', the building standpipe system would have been extended to three zones.

In the two-zone example, the lower zone would have been a standard standpipe arrangement, with the possibility of a gravity tank at the top of the zone supplying the lower zone standpipes. (See Figure 3.15.) The upper zone standpipes would be fed by express risers or risers that do not supply water to any system below the base of the upper zone standpipes. The upper zone standpipes may also have the additional supply of a gravity tank.

Due to the excessive height of both of these zones, a fire pump will undoubtedly be required to provide sufficient pressure to overcome the static (elevation) pressure loss and still furnish 65 pounds at the topmost standpipe outlet. In the example, the topmost outlet of the upper level is 355' above the supply, and it would take 154 psi just to raise the water to this height. This is one reason for the gravity tank above the standpipe zone elevation to maintain positive pressure on the system.

Each zone of high rise zoned standpipe systems requires a separate siamese connection. According to conditions, one fire pump may supply the lower zone in the example and the lower zone gravity tank, and the second fire pump located at the bottom of the upper zone takes its suction from the lower zone gravity tank. This arrangement prevents the upper zone from having to overcome an additional 87 psi static pressure to get the water to this elevation.

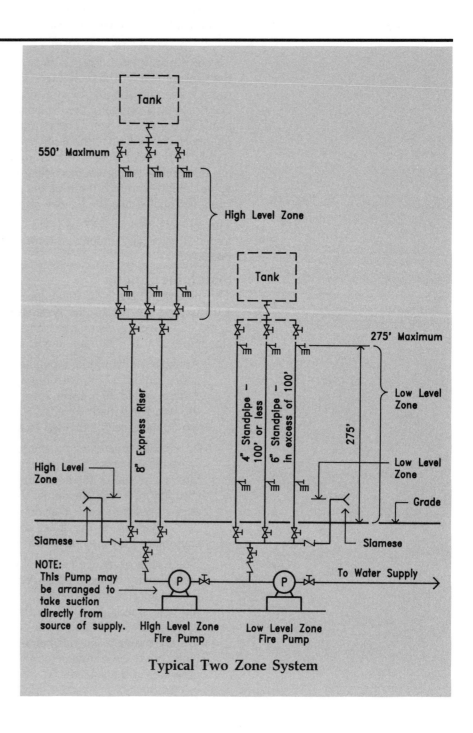

Typical Two Zone System

Figure 3.15

For multiple zones, additional pumps for the upper zones will be required. These pumps are located at the base of the higher zones. The lower pumps supply water to the gravity tanks, and the upper zone pumps take suction from the gravity tanks.

Combined Standpipe and Sprinkler Riser

Until a few years ago, standpipes were an expensive addition in some structures, especially in high rise buildings. It was necessary to run a riser pipe for the sprinkler systems and a riser pipe for the standpipe systems. New regulations allow a single riser pipe to supply both sprinklers, hose stations, and 2-1/2" hose outlets. Additional equipment is required for each floor's sprinkler supply coming from the riser such as a control valve, drain, and test connection.

The combination sprinkler and standpipe riser installed in the stairtower supplies the 2-1/2" fire department valve at each floor level. If 1-1/2" hose stations are required—although they are usually not required when the building is totally sprinklered—the supply for these hose stations can be piped from the riser throughout the area.

Water Supply

The supply for sprinklers is obtained from the riser at each floor. This supply to sprinklers at each floor must be controlled by an *indicating type control valve*.

Should the sprinkler system on one floor of the building have to be repaired or remodeled, it will only be necessary to shut down this one floor system, leaving the balance of the sprinkler and standpipe systems in service. With the water supply valve shut, but the piping full of water, it is necessary to provide a drain valve on the system side of the floor control valve. This drain valve must be connected to a drain line to discharge the system water to a safe location.

In most high rise buildings a *sprinkler water flow alarm* is necessary at each floor so the fire department can locate the floor where the fire has activated sprinkler heads. Each sprinkler supply from the riser at each floor is provided with a *paddle-type flow switch* on the system side of each floor control valve. Now it is also necessary to test the individual flow switches and, in addition to the drain, a test valve must be provided.

The combination drain and test line with drain and test valves can be piped as follows: A restricted orifice giving the flow access to one sprinkler head to operate the flow switch must be installed in the test line. There are devices available that combine the drain and test in one single element, and these can minimize the space required to individually pipe the drain and test.

Standpipe systems provide adequate equipment to combat fires in different stages of growth. The small hose stations give occupants the maximum opportunity to provide help in a fire emergency, and 2-1/2" outlets provide firefighters with effective equipment to combat a fire that has reached major proportions. Standpipe systems must be designed based on a complete study and evaluation of all facets of the structure, occupancy, hazards, and personnel. Standpipe systems are not a substitute for sprinkler systems, but they can be a tremendous complement to automatic protection systems.

Water Spray System

In discussing deluge systems, the sprinkler heads are open, but there are numerous instances requiring water discharge over an irregular mass, like a transformer, or a discharge of very fine water droplets in order to provide more cooling or to afford maximum cooling on a surface exposed to the radiant heat of a fire. The discharge pattern of a sprinkler head may be too uniform, and the water droplets too large to accomplish this protection, so the sprinkler heads are replaced with water spray or nozzles (water fog nozzles). Although water fog is a trade name, it is sometimes used to describe water spray systems, regardless of the manufacturer. Since the current sprinkler heads are referred to as spray sprinklers, the term water fog has gained even greater general use to identify true water spray systems.

A water spray system is a deluge system using specially designed nozzles (see Figure 3.16) in lieu of open sprinkler heads on a deluge system. Being a deluge system, the deluge valve is activated by a detection system. When the deluge valve is activated by a signal from the detection system, water is discharged from all nozzles on the system simultaneously.

Unlike the requirements for sprinkler heads, water spray nozzles can be spaced, located, and positioned in any configuration that provides effective coverage of the protected area. The water spray pattern, velocity, and density determines the location and positioning of the water spray nozzles.

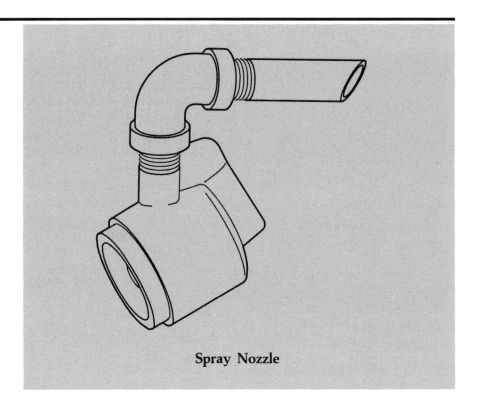

Spray Nozzle

Figure 3.16

Determination of the proper spray pattern and the positioning of the nozzles to take full advantage of the spray pattern requires experience, expertise, and professional judgment. Each water spray system is individually designed, "tailor made," for protection of a specific hazard. NFPA publishes a Fire Code (NFPA Standard 15) that furnishes technical requirements for the design and engineering of water spray systems "based upon sound engineering principles, test data, and field experience." But the selection of nozzle patterns, velocity of discharge, density, positioning of nozzles, etc., is the responsibility of an experienced professional.

Some hazards to be protected with water spray are so unique that a fire protection company will construct a "mock-up" of the equipment, using the actual flammable liquid involved at their testing site. The test data obtained is then used to design and engineer the water spray system for the hazard.

There is a very thin line separating water spray and standard sprinkler protection: A water spray nozzle and sprinkler head discharge differs only in the discharge *pattern*, *velocity*, *particle size*, and *density*. Water spray systems may be installed to satisfy several fire protection situations; extinguishment, controlled burning (limitation of fire spread), exposure protection, and fire prevention. Although a water spray system may be designed to perform all of these functions, they will be evaluated individually.

Extinguishment

In the history of sprinkler head development, one of the primary fire extinguishing advantages of the spray sprinkler head over the old-style sprinkler head was its ability to break up the water discharge into smaller particles for better cooling and smothering action when the droplets were converted to steam. Water spray system nozzles produce a water spray consisting of very small water particles that are easily converted to steam. This action not only takes heat from the combustion in changing the liquid to vapor, thus cooling the combustion, but also the steam produced keeps oxygen from reaching the fire and produces a positive smothering action.

Water spray systems have been used, to a limited degree, to extinguish certain flammable liquid fires by *emulsification*, *dilution*, or both. Emulsification is the mixing or combining of the flammable liquid and water to form an *emulsion*: In other words, a state of suspension of water droplets in the flammable liquid, similar to fat in milk. With emulsion extinguishment of oil fires, the flames in the vapor-air mixture above the surface of the burning liquid are diluted by the water vapor of the water spray to a point where there is no longer a flammable vapor-air mixture and the fire is extinguished.

Some water spray droplets are vaporized as they pass through the heat of the flames, but the primary source of effective water vapor comes from the water film in the emulsion at the surface of the burning liquid. The film is vaporized by the heat of the burning liquid, and this vaporization process reduces the temperature of the burning liquid. The vaporization of water in the emulsion also reduces the volume of flammable vapors and produces a barrier on the surface which excludes air (oxygen) from the combustion.

Vapor dilution, the water film in emulsion at the surface of the burning liquid, can only be accomplished if droplets in the water spray are large enough to pass through the flame area without being vaporized and reach the surface of the liquid. The flammable oil vapors immediately above the surface are too rich to burn until they rise and mix with air to form a combustible mixture. The vaporizing emulsion of water and flammable liquid at the surface in this rich vapor area produces a mixture of water and oil vapor that will not support combustion, and the fire is extinguished.

Lighter liquids, like kerosene, do not form an emulsion with sufficient stability to cause extinguishment by this method. To extinguish these lighter liquids and oils, the vaporization of the water spray droplets as they pass through the flames produces a cooling and smothering extinguishment.

Dilution is applying sufficient water to dilute the flammable liquid to the point where it cannot produce sufficient flammable vapors to support combustion. Dilution of a flammable liquid usually requires so much water that it becomes an impractical fire protection consideration.

Controlled Burning

The deluge concept of a water spray system, coupled with the cooling action of the water spray, can effectively limit the spread of a fire, even if the fire is of such intensity that the water spray system cannot complete extinguishment. In some specific instances a burning material cannot be extinguished regardless of the amount of water applied, and the intent of the water spray system is merely to confine the combustion to the point of origin.

Another function of controlled burning is confinement of a fire that should not be extinguished, for example, a fire at a gas line leak. As long as the fire continues to burn, the escaping gas is consumed. But if the fire were extinguished, the escaping gas could spread and create a life safety situation or collect in a low point some distance from the leak, make contact with an ignition source, and ignite or explode. The water spray system will be designed to keep the surrounding piping and equipment cooled to prevent explosions, fires, or damage from the radiant heat. The water spray discharge will also keep the fire controlled to reduce the intensity of the radiant heat produced.

Exposure Protection

Many water spray systems are designed specifically to remove or reduce the heat from the surface of structures or equipment exposed to the radiant heat of a fire. A research firm recently conducted tests to determine the effectiveness of deluge systems with standard sprinkler heads on aircraft jet fuel spill fires. The data from these tests accents the value of water spray structural steel exposure protection. For example:

> One test consisted of a fuel spill of 900 to 1300 square feet and involving several hundred gallons of jet fuel. The very turbulent fire plumes that were formed by this fire attained velocities in excess of 50 miles per hour and steel specimens, 6 × 8 I-beams located at a height of 40' above the fire, and at the ceiling 60' above the fire, attained temperatures between 800°F and 1200°F within a very few minutes.

At such temperature levels, structural steel under a load condition will deform. When steel deforms, severe structural damage can occur, therefore, the exposure protection of a water spray system can easily be justified when a steel structure or structural steel supports are exposed to an intense fire condition.

Designs have been prepared using the water spray system as a *water curtain* separating the fire area and the equipment to be protected. Water spray curtains can be effective under certain conditions, but direct application of the water spray is much more efficient exposure protection.

Fire Protection

An example best illustrates this function of a water spray system:

A bank of three large oil-cooled transformers at a power generating facility each contain 10,000 gallons of oil. An oil reservoir on one transformer is ruptured and oil starts to spread under the other transformers. Every hot surface or high voltage electric contact presents an ignition source. The water spray systems surrounding each transformer are actuated by tripping the water spray deluge valves automatically by a detection system. The water spray will disperse the oil and, to a degree, dilute and thin the oil spill, which very effectively keeps all surfaces cool, preventing ignition. Should one transformer become ignited, water spray on the uninvolved transformers will prevent or control radiant heat damage.

Using transformers as an example serves two purposes: (1) to point out the use of water spray systems as a fire prevention system, and (2) the protection of oil-filled transformers with water spray. An electrical short in a high voltage transformer can create enough heat to vaporize the oil. The resultant pressure can rupture the oil reservoir casing, spewing hot oil over hot transformer surfaces. The resultant fire is extremely intense, and exposure protection with water spray of the other uninvolved transformers becomes of the utmost importance.

Electrical Conductivity

Water spray nozzles placed at a predetermined distance from electrical equipment will *not* conduct an electrical current on the water droplet discharge. This is a major consideration in fighting a transformer fire, or a fire in any high voltage electrical equipment. A solid water stream or the relatively heavy and large water droplets produced by sprinkler head discharge can easily conduct an electrical current through the water and back through the piping, threatening life safety.

Water Spray vs. Sprinkler Systems

Before examining the design features of water spray systems, it is important to clarify the interface between sprinkler and water spray systems. Water spray systems are *not intended to replace* automatic sprinkler systems, although water spray systems may, under certain circumstances, supplement or compliment sprinkler systems in protecting special hazards.

Design Considerations

The first determination in recommending a water spray system are (1) the nature of the hazard, and (2) the specific purpose for using this system. The hazards most frequently protected with water spray systems are those involving flammable liquids, gas tankage, associated piping, devices, and equipment, electrical equipment such as transformers and oil switches, generators, and other rotating electrical machinery. Primary design considerations for a water spray system are the anticipated heat produced by the combustion, and the radiant heat from exposure fire. Once this is determined, a density (gpm per square foot) sufficient to absorb the combustion and exposure head can be calculated.

Nozzle Selection

Another consideration is the *droplet size* produced by the nozzles to determine if the water spray should be coarse or fine. The extinguishment of a fire occurs at an interface between the surface of the burning material and the base of the flame, so it is necessary to use a nozzle that will not only produce a water spray at the required density, but also a water spray delivered at a velocity and mass (droplet size) to penetrate the heated air currents and reach the fire zone. In the case of exposure protection, the surface of the equipment must be completely covered in spite of heat waves radiating off the metal surface.

Nozzles for water spray systems come in all sizes and shapes with an equal multitude of discharge patterns and methods of developing the spray pattern. The engineer must select the proper nozzle to fit the conditions: the nature of the hazard, the purpose of the system, and in severe wind or draft conditions, the deflection of the water spray.

Generally, high velocity nozzles produce a cone pattern, and low velocity nozzles deliver a sheroid or cone pattern consisting of a much finer spray. With most water spray nozzles, the higher the velocity, the coarser the spray or water droplets, and the greater the range of reach of the spray.

On the market there are several types of very high velocity, high capacity, spray nozzles designed primarily for transformer protection. With heavy capacity very few of these nozzles are required. The transformer can be removed without disturbing the water spray system piping because the nozzles can be located at some distance from the transformer, with wide spacing between nozzles.

Water spray nozzles produce their characteristic spray pattern in different ways depending on the manufacturer. One spray nozzle develops a high rotary motion of the water inside the nozzle head by the use of spiral passages, another by impingement of small, carefully directed jets of water through opposite multiple discharge openings in the nozzle body, while a third type impinges small jet streams of water by multiple small opposed openings on the outside of the nozzle body.

Spray nozzles are available that not only resemble regular pendent sprinklers, but also function as a regular sprinkler—the only difference is the specially designed deflector. These spray nozzles may have a fusible link similar to a standard sprinkler head.

There are many other approved spray nozzles on the market, but one unique nozzle deserves mention. The *spiral-type spray nozzle* is designed to discharge water along the axis of an external spiral with a diminishing inside diameter. The water discharging over and around this spiral literally peels off in thin layers from the surface of the cone, and this thin layer of water breaks into a spray as it leaves the spiral.

It is usually a requirement of the insurance interests, and is always a recommendation of the fire protection engineer, to install a strainer in the main water supply to a water spray system to remove any foreign material such as sand, dirt, or scale before it lodges in the small orifice of a spray nozzle and disrupts the discharge pattern. Water spray nozzles that depend on very small jet stream orifices to create a specific discharge pattern will have their own individual internal strainer at each nozzle, but this does not eliminate the need for the main line strainer to remove larger foreign material before it can clog the nozzle strainer. Main line strainers are designed so they can be flushed periodically.

In addition to the possibility of internal clogging of small orifices, spray nozzles installed in an atmosphere similar to paint vapor or a corrosive atmosphere may require blow off caps. Blow off caps fit over the nozzle in such a manner that the water pressure will immediately blow them off. To prevent personnel from being struck by the blown off cap, it is attached by a chain or cable to the nozzle body. This attachment allows for reuse after the system is shut off.

Configuration of Equipment

Once a nozzle is selected, the next step is to study the physical arrangement and configuration of the structure or equipment to be protected. This criteria will enable the engineer to determine the proper placement of the nozzles and the angle at which the nozzles will be installed, since each nozzle has a predetermined discharge cone or pattern. The nozzle discharge capacity, location, and spacing of the nozzles, angle of discharge, and the nozzle discharge cone or pattern, all contribute to the delivery of the desired density requirement over the entire surface to be protected.

Water Supply

Hydraulic calculations based on supply and demand will determine the proper size and water requirements. Water supply needs must take into consideration the number of nozzles that will operate in any anticipated fire condition, and the required duration of the water supply to the water spray system, and this must be evaluated for each installation. The anticipated duration of discharge depends on the nature of the hazard and the purpose for which the system is designed.

In discussing the water supply, two factors warrant discussion. First, the *number of nozzles* that will operate. Most water spray systems are *open head deluge type*, but closed orifice nozzles are available. These systems operate like sprinkler systems, with only the involved nozzles operating. Secondly, the *size of the system*. Insurance standards recommend that the size of water spray systems be limited so that the designed discharge rate, calculated at the minimum pressures required for the nozzle discharge to be effective, will not exceed 3,000 gpm.

Sizing the System

Naturally, the size of a water spray system is governed by:

- The nature of the hazard
- The combustibles involved
- Size of the area being protected
- Type and amount of equipment or structural elements being protected
- The size of the potential fire hazard

Other systems in the area could limit the size or spread of a fire. Fire wall subdivision of the area would afford containment of a fire, and dikes, curbs, and drainage facilities can limit the spread and extent of a flammable liquid fire.

Flow from a water spray system can vary from a few hundred to several thousand gallons per minute depending on the type of hazard and the extent of the anticipated fire area. The duration of the flow from a water spray system depends on many factors and can vary from two minutes to more than two hours when exposure protection is involved.

Water spray systems usually protect high value and high hazard equipment or facilities, and the detection system that activates these systems must be reliable, dependable, and in some cases, quite sensitive.

Foam Systems

In August 27, 1859, Edwin L. Drake struck oil 69-1/2' below the earth's surface near Titusville, Pennsylvania, and created the first producing oil well in the United States, pumping 25 barrels a day. The discovery ushered in not only a new era in technology, but also the beginning of new and unique fire protection problems.

The oil era actually began in 1855, four years before Mr. Drake struck it rich in Pennsylvania, when a chemist named Benjamin Stillman made the first fractional distillation of oil. Stillman's process was initially used to obtain oil for lamps, but this led to the discovery of higher boiling fractions which were useful as lubricants. Later, a lower fraction gasoline would find its application in powering the automobile.

Between the Civil War and the turn of the 20th century, the chemical industry made tremendous strides and mushroomed into a giant among business enterprises. The fire protection industry, accordingly, must keep up with the growth of new technology. New products, new processes, and new equipment must be developed to control new fire hazards, and this was especially true of the flammable liquids that were developed.

America really entered the liquid fuel age at 10:30 a.m. on the 10th of January in 1901 when "Spindletop" roared like a cannon and spouted oil a 100' over the top of the derrick. Spindletop was located in Texas, and after drilling through 140' of solid rock to a total depth of 1020', Spindletop spewed 800,000 barrels of oil in nine days before it was capped. Spindletop yielded more oil in the first year than the entire world was producing at the time.

More important than the statistics of this first oil well in Texas was the fact that Spindletop revealed to the world the nation's greatest oil field, ushered in the modern petroleum industry, and opened the door to the liquid fuel age. Oil discovery in Texas revolutionized industry and raised the nation's standard of living.

Flammable liquid fire protection during the years following the success of Spindletop was literally a cannon maintained in a loaded and ready condition. A fire in a holding tank of oil resulted in the firing of the cannon to shoot holes in the tank or blow it apart so the fuel could fill the dike where it could burn itself out.

In 1896, there were four operable automobiles in the United States, and by 1911 there were more than 600,000. In 1903, Orville Wright successfully piloted the "Flyer" at Kitty Hawk, and the age of aviation was born. The value of petroleum and its usefulness increased as automobiles and airplanes developed and multiplied. The other uses for petroleum rapidly made this resource the starting point for thousands of hydrocarbon-based products.

Naturally, flammable liquid fires were common, particularly when the internal combustion engine became practical, creating a tremendous market for gasoline. Water was applied to these flammable liquid fires, but this traditional firefighting agent was not very effective in control or extinguishment. As always, scientists saw a challenge, industry saw a market, and the result was the development of a new flammable liquid fire extinguishing agent—foam. When a flammable liquid burns, it is the vapors from the surface of the liquid that are burning.

The lowest temperature of a flammable liquid that will produce vapors in enough concentration to form a mixture with air to ignite the vapors at the surface of the liquid is the *flash point* of the liquid. The flash point of a liquid determines the classification of the liquid as a flammable or combustible liquid. Each of these liquid classifications is subdivided into subclassifications by temperature range and by the boiling point of the liquid, but the basic, conversational, difference is this: A flash point below 100°F classifies the liquid as a flammable liquid, whereas a flash point at or above 100°F classifies the liquid as a combustible liquid.

This generic classification of flammable and combustible liquids is extremely important when considering an adequate protection agent and system. Consider how easily the fumes from low flash point gasoline can be ignited at room temperature, and how difficult it is to ignite the fumes from fuel oil at room temperature.

Since the vapors given off by the liquid ignite and burn, the best way to extinguish a flammable or combustible liquid fire is to cut off the supply of vapors, to prevent air from mixing with the vapors, or both.

Foam is one answer. Foam applied to a burning liquid fire flows over the surface of the liquid and forms a blanket that not only prevents the vapors from leaving the surface of the liquid, but also prohibits the air from reaching the vapors to form an ignitable mixture of vapors and air. When there is a flammable liquid fuel spill which has not ignited, foam applied to the spill will prevent the vapors from leaving the surface and air from mixing with the vapors, should the spill contact an ignition source. This is foam application in the role of fire prevention.

Foam is an aqueous (water) suspension of air or gas in the form of small bubbles surrounded by a film of solution. Foam is lighter than the flammable or combustible liquids so the blanket of foam will float on the surface of the liquid. In addition, there is a certain measure of cooling action by heat absorption from vaporization of the water contained in the very thin films in the foam. This heat absorption is both derived from the fuel surface and adjacent metal surfaces. Some theories consider this cooling action to be a major factor in the extinguishing action of foam, and in some applications, with some types of foam, this may be true.

A good foam blanket will strongly resist the heat and flame of the fire, as well as disruption due to wind, thermal drafts, and mechanical agitation, and will reseal an opening in the blanket. One modern foam (aqueous film forming foam) does not have this foam blanket integrity, but it is far superior to other foams in its ability to accomplish a quick "knock-down"—extinguishment—of a flammable liquid fuel spill fire.

Use and Calculation of Foam
Foam is usually used with open head deluge systems, but with careful attention to proportioning (explained in the next section), foam systems can be used with wet pipe, dry pipe, and pre-action systems.

With deluge systems every head is discharging, and knowing how much gpm each head is discharging at known pressure, the demand, or total gpm, is easily calculated. Knowing that the demand will be constant, proportioning the foam concentrate into the water supply can be calculated based on this constant demand.

With a closed head wet pipe, dry pipe, or pre-action system, the demand cannot be accurately determined because the combustion may fuse two heads, 10 heads, or 100 heads. The demand will be based on the number of heads that fuse and discharge foam/water solution. Attempting to accurately proportion the correct percentage of foam concentrate into the water at different flows is an extremely difficult design situation requiring professional experience and expertise.

Some foam deluge systems use spray heads as discharge devices. The discharge pattern from these foam spray nozzles is similar to the sprinkler and foam/water sprinklers that have been described, but these foam spray nozzles are larger and have a heavier discharge.

Proportioning

Basically, foam concentrate is mixed with water to form a foam/water solution which is transported through a piping system to a discharge device or devices where it is mixed with air (aerated) to produce the foam.

There are any number of methods of injecting the foam concentrate into the water in a predetermined proportion. The equipment or device to accomplish this action is the *proportioning* part of a foam system.

Foam concentrate is available in a number of concentrations, but the most common are 3% and 6%. When 3% foam concentrate is introduced into the water in the system, the solution that is transported to the discharge device or devices consists of 3% foam concentrate and 97% water. The same proportions are created with 6% foam concentrate, namely 6% foam concentrate and 94% water.

Correct foam proportioning is one of the most important factors in the design of a foam system because the quality of the foam produced depends on the efficiency of the proportioning system. The optimum quantity and quality of foam discharged on a flammable liquid fire must be obtained to produce the desired extinguishing results.

Should the percentage of foam concentrate in the water solution be too high, the discharged foam will be too thick and will not flow quickly across the surface of the flammable liquid or around obstructions. In addition, because more foam concentrate is being proportioned into the water than required, the supply of foam concentrate will be depleted too quickly.

Should the foam concentrate be proportioned into the water at less than the required percentage, the discharged foam solution will produce a foam blanket that has less resistance to break down from the heat and flame of the combustion. The water in the foam will drain off faster and cause quicker foam blanket deterioration.

The foam solution is transported through the system piping until it reaches the *discharge devices*, and these devices perform the function of mixing the foam solution with air to form the characteristic foam blanket. The discharge devices are often regular automatic sprinkler heads. Since the foam solution is aerated, or mixed with air, by being discharged against the deflector of the sprinkler head without any special device other than the deflector to aerate the foam solution, these sprinkler heads are referred to as *non-air-aspirating foam sprinklers*. (See Figures 3.17a and 3.17b.)

There are other foam discharge devices that aerate the foam solution within the shaft of the head. The discharge from these heads against the deflector creates the discharge pattern. These heads are *air-aspirating foam discharge heads*.

Both the regular sprinkler heads, non-air-aspirating heads, and the air-aspirating sprinkler heads can function as regular sprinkler heads when the foam supply is exhausted. The discharge of water from these heads will produce the typical sprinkler head spray pattern.

Aqueous film forming foam (AFFF) is a much more effective extinguishing agent on flammable liquid combustion when discharged through regular, non-air-aspirating, sprinkler heads. An example is the lower density (gpm/square foot) required when using AFFF discharged through regular 1/2" orifice non-air-aspirating sprinklers on a deluge system protecting an aircraft hangar against jet fuel spill fires. The density with these sprinkler heads is 0.16 gpm per square foot with AFFF, and 0.20 gpm per square foot with air-aspirating foam/water heads.

Foam-Aspirated vs. Non-Aspirated Discharge Devices

(*courtesy of Ansul Fire Protection*)

Figure 3.17a

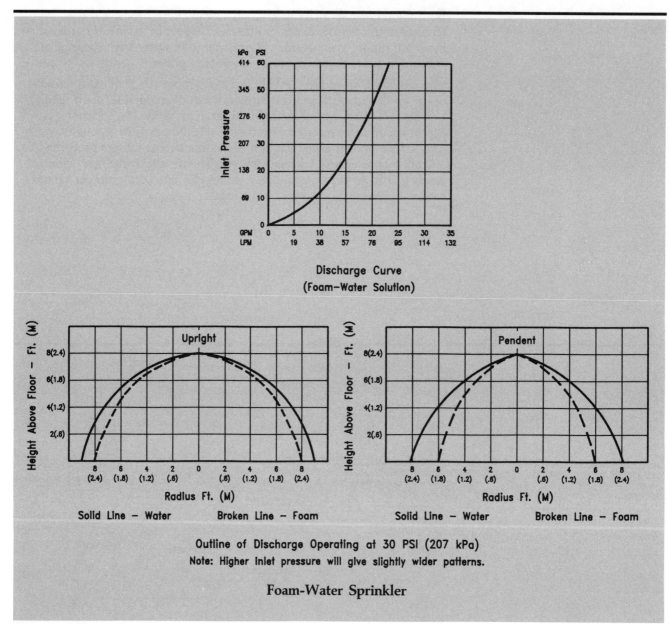

Foam-Water Sprinkler

Figure 3.17b

Selecting Proportioning Systems and Devices

Efficient foam concentrate proportioning is required to provide quality foam production that insures optimum performance as an extinguishing agent. Proportioning is the heart of a foam system. As with the results obtained from a computer, the results of a foam system are only as good as the material fed into it. A foam system is only as effective as the proportioning system that introduces the percentage of foam concentrate into the system.

The required foam/water solution flow rate to meet the design criteria and the available water pressure are the two most important features in selecting the proper proportioning arrangement.

There are certain nationally specified amounts. Underwriters Laboratories (UL) requires the following proportioning range for proportioning devices. These are the limits allowed when using 3% or 6% foam concentrate:

3%—not below 3% or above 4%
6%—not below 5% or above 7%

Even though most proportioning systems and devices are shelf-items, it requires experience and expertise to select proportioning that satisfies the overall situation. Each proportioning system or device has its unique advantages and disadvantages, depending on the specific system, system application, and conditions relative to the hazard.

If the shelf-item proportioners cannot be used, a proportioning system design must be specifically developed to accommodate special circumstances and conditions.

Familiarization with several of the standard proportioning systems and devices is important in any discussion of foam system protection.

Balanced Pressure Proportioning

The balanced pressure proportioning system uses a tank to hold the supply of foam concentrate. The *atmospheric foam concentrate storage tank* is not a pressure tank, since the foam concentrate does not have to be under pressure with a balanced pressure proportioning system.

The tank can be constructed of mild steel, fiberglass reinforced plastic, or polyethylene plastic. The foam concentrate tank material must be compatible with the type of foam concentrate to be stored in the tank. There are many atmospheric storage tanks on the market, but specifying or purchasing one that is not specifically designed and constructed for the storage of foam concentrate will not only lead to expensive system problems, but will reduce the system's suppression effectiveness. The concentrate life will be shortened, but more important, the concentrate could become contaminated and not produce effective foam for fire suppression.

In this system, the foam concentrate is pumped from the storage tank to the proportioning system. The foam concentrate pumps are *positive displacement pumps*, not centrifugal pumps. The foam concentrate pump size depends on the design rate of flow (gpm) of the concentrate and the type of foam concentrate. With positive displacement pumps, the discharge volume at a specified pressure is constant. In the chapter on pumps, it will become obvious that a positive displacement pump discharges on a straight line pump—constant volume at constant pressure. A centrifugal pump discharge curve indicates an increase in discharge volume as the discharge pressure decreases.

The foam concentrate is pumped through a bypass from the main concentrate discharge line to a device called an *automatic pressure balancing valve*. This valve, which is also piped into the main water supply line, regulates the foam concentrate pressure to match the water pressure. As the water pressure varies, the pressure balancing valve adjusts the foam concentrate pressure to maintain the required flow of concentrate into the water.

The water supply may be supplying several deluge systems, and as the combustion operates more systems the water pressure will decrease as the volume demand increases. The pressure balancing valve automatically reacts to this pressure change and maintains the flow of foam concentrate consistent with the changing water pressure. Excess foam concentrate above the required design flow is returned to the storage tank.

The foam concentrate discharge line from the pump is connected to a device called a *ratio controller*, or *concentrate metering device*, installed in the main water supply line. This device is a carefully designed *venturi*. (Recalling physics, the water passes through a throat in the venturi. Since the volume of water into the device is the same volume out of the device, the velocity of the water increases as it rushes through the narrow throat. When the velocity increases, the pressure decreases). The foam concentrate is under the same pressure of the water supply controlled by the pressure balancing valve, and when the water pressure decreases in the narrow venturi throat, foam concentrate is forced into the water stream.

With the pressure balancing valve adjusting the foam concentrate pressure to match the water pressure, and the drop in pressure at the throat of the controller, the foam/water solution leaves the outlet of the controller at the correct foam concentrate-to-water percentage, and the foam/water solution is piped to the system discharge devices.

Reserve Systems
On projects where the foam system is protecting high values, or when the insurance carrier requires maximum, uninterrupted system protection, two foam concentrate storage tanks and two foam concentrate pumps will be required. The second reserve tank is permanently piped into the system, and by manual operation of a valve it will become the primary supply when the first tank has exhausted its supply. Should one pump be out of service, the system is still in a functional condition.

Direct Orifice Proportioning
If there are long runs of pipe from the proportioning system to the foam system piping and discharge devices, the transit time of the foam solution may be several seconds. Some flammable liquid fire hazards cannot afford the luxury of waiting several seconds for the foam discharge to suppress and extinguish the combustion. Aircraft hangars are an example.

A jet fuel spill fire beneath a jet aircraft can cause irreparable damage to the aircraft in the hangar if the combustion is not controlled in 30 seconds and extinguished in 60 seconds. So allowing several seconds for detection activation, the discharge time of the foam system becomes a few critical few seconds.

Foam deluge systems that protect this type of hazard have a deluge valve located as close as possible to the system that they supply. Water under pressure is supplied to each deluge valve, and foam concentrate is supplied to a small foam concentrate deluge valve installed immediately adjacent to the water deluge valve. The foam concentrate that supplies the foam concentrate deluge valve is maintained at a higher pressure than the water supply to the water deluge valve. The outlet pipe of the foam concentrate deluge valve is piped into the riser of the water deluge valve above the deluge valve. In this foam concentrate line is a carefully calculated *orifice plate*. The orifice plate has a small hole in the center to allow the flow of foam concentrate. When the foam deluge system detection system is activated, both the water deluge and foam concentrate deluge valves trip simultaneously and foam concentrate is forced through the orifice and into the water flow leaving the water deluge valve. The foam concentrate pressure is higher (by approximately 15 psi) than the system water pressure in the riser and can therefore be forced into the water stream in the system riser. Based on water and foam concentrate pressure at the two deluge valves, the orifice size is calculated to admit the proper percentage of concentrate into the water.

A device detects the instant drop in pressure on the foam concentrate supply line and water supply line when the deluge valves trip, and this pressure-sensitive device automatically starts the foam concentrate and fire water pump.

In the chapter covering pumps it will be noted that some large diesel drive fire (water) pumps require several seconds to come up to speed. The diesel engine must run at full power so the connected fire water pump can discharge at full volume and pressure.

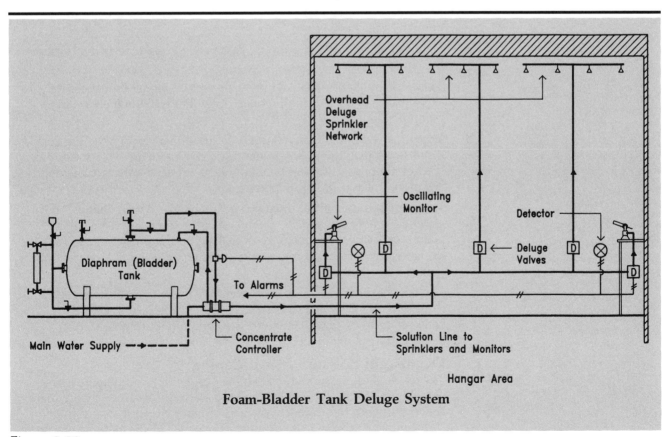

Foam-Bladder Tank Deluge System

Figure 3.18

When a hazard requires the discharge of foam without delay, the foam concentrate pump and fire water pump can be started electrically when the first signal is transmitted from the detection system. The pressure-drop pump-starting system is then merely a back-up.

Detection Systems

Foam deluge systems are usually designed to prevent a false "dump" of the system when the detection system is activated by some non-fire condition or the combustion is large enough to activate a detector, but small enough to be extinguished by hoses or extinguishers.

The detection system is arranged so that the first detector to be activated activates all alarms, and the second detector to be activated dumps the deluge systems. To accomplish the starting of the water and foam pumps electrically, the water and foam pump starting sequence would take place when the first detector is activated.

The water fire pump will harmlessly churn, and the foam pump will discharge foam concentrate back to the tank. If the combustion develops and intensifies sufficiently to activate a second detector, water, foam pressure, and volume will be immediately available.

Jockey Pump

In order to maintain the water supply and foam concentrate supply to their respective deluge valves under pressure, both the water line and foam concentrate line will have a small gpm pressure maintenance pump—commonly referred to as a *jockey pump*.

The water jockey pump takes suction on the supply or suction side of the fire water pump and discharges into the supply line on the discharge side of the fire pump check valve. The pressure is maintained in the supply line to the deluge valve by being blocked from passing back through the pump by the check valve.

Here, too, the concentrate cannot pass back through the pump because of the check valve.

These jockey pumps maintain pressure slightly higher (approximately 10 or 15 psi) than the starting pressure of the fire water and concentrate pump pressure switch.

When the water and foam concentrate deluge valves trip, the pressure drops faster than the jockey pump can compensate for. The pressure drop reaches the pre-set starting pressure of the pressure switch of the fire water and foam concentrate pumps.

Jockey pumps are further discussed in Chapter 5, "Water Pumps," but water and concentrate jockey pumps should be small volume (gpm) pumps so there is no possibility of making up the pressure loss when the deluge systems trip. This would prevent the fire water and concentrate pumps from starting.

The specifications for the foam concentrate jockey pump should specify that the pump parts and construction shall be compatible with the concentrate being used to prevent the concentrate from rapid deterioration of the pump, gaskets, seals, and metal parts.

Diaphragm Pressure Proportioning

The diaphragm pressure proportioning system does not require foam concentrate pumps because it supplies water pressure to discharge foam concentrate into the water supply and produce the foam/water solution. (See Figures 3.18 and 3.19.) The *concentrate tank* is a metal pressure vessel containing a collapsible bladder, and thus it derives the name of *bladder tank*. This foam system is usually referred to as a *bladder tank system*. The bladder contains the concentrate, and when water pressure is supplied to the tank surrounding the "full" bladder the water pressure squeezes the bladder and forces the foam into the piping.

Water Supply

The water supply line is equipped with a *ratio controller venturi*, and on the water supply side of the venturi a small water line is connected to the water supply side of the bladder tank. The pressure on the outside of the bladder is thereby equal to the water supply pressure.

The concentrate bladder has a concentrate line from the bladder to the throat of the venturi of the ratio controller. This line is kept closed by an *automatic valve*. When the water supply valve is tripped by the system detection system, the valve on the concentrate line to the venturi is opened simultaneously.

Foam-Bladder Tank Closed Head Sprinkler System

Figure 3.19

Supply water pressure is transmitted to the outside of the bladder in the tank, and foam concentrate is pushed through the concentrate line into the venturi. Here it is proportioned into the water, and the foam/water solution is piped to the system discharge devices. The pressure of the concentrate is equal to the pressure of the water and it can be injected into the water supply where the water pressure drops as the velocity increases at the throat of the venturi.

The atmospheric tank on the balanced pressure proportioning system can be refilled with concentrate during operation of the system, but a bladder tank is a pressure vessel and therefore cannot be refilled during system operation. Refilling the bladder tank with concentrate requires considerable time and care to avoid pinching a hole in the bladder. Even a pin hole in the nylon reinforced elastomeric bladder will render the bladder tank useless due to loss of pressure necessary to force concentrate out of the bladder.

Some bladder tanks have a diaphragm instead of a bladder that completely fills the tank. These tanks are flanged in the center and the internal diaphragm is fastened to this flange. When water pressure is applied to the diaphragm, which is at one end of the tank, the concentrate is forced through a concentrate line at the opposite end of the tank and thus to the ratio-controller venturi. (See Figure 3.20.)

High Expansion Foam

The foam that has been previously discussed is called *low expansion foam*. When low expansion foam is discharged, it forms a layer of foam about 2"-3" deep after a 10-minute discharge from a deluge system. In contrast, the discharge of a high expansion foam system for 10 minutes can form a layer of foam 30' high. The expansion ratio of low expansion foam is about 8:1 or 10:1. High expansion foam expansion ratio is about 500:1 or 1000:1.

High Expansion Foam Generators

To understand how high expansion foam is formed and discharged, recall blowing bubbles through the film on a plastic circle. One bubble of high expansion foam is similarly produced. A 1%–3% foam concentrate is proportioned into the water, and the foam/water solution is sprayed over a metal or nylon mesh. Air passes through the mesh at a high velocity, creating the high expansion foam discharge. High expansion foam consists of millions of tiny individual bubbles. The foam may be as much as 1000 parts air to one part water. In contact with the fire, the foam changes to steam (which expands 1,700 times) and the oxygen content of this steam and air mixture becomes less than 7.5%. The combustion is oxygen-starved and extinguished, literally suffocated.

High expansion discharge devices are called *high expansion foam generators*. These foam generators are usually installed on the roof or high on the walls so the foam will flow into the protected area. Unlike low expansion foam discharged from sprinklers or nozzles under pressure, the only pressure on the high expansion foam comes from the air pressure through the liquid foam solution on the mesh or screen. By mounting the foam generators at a high point, the discharge can use gravity to flow into the hazard area.

High Expansion Foam Fire Suppression

High expansion foam also assists in controlling and extinguishing combustion by diverting the flow of air into the fire area by the mass of the high expansion foam discharge. The heat from the combustion forms a thermal column that sucks air into the fire area. The high expansion foam discharge, by its volume, disrupts this flow of air.

In addition, high expansion foam has a cooling action because of the discharge water content. This water held in suspension cools the surface of everything it comes in contact with in the hazard area.

The bubbles in high expansion foam are relatively uniform in size, the walls are quite tough and elastic films, and there is very little liquid trapped in the bubble to weaken and distort its configuration. All these elements produce a foam with very stable liquid retention, important to cooling action and extinguishment.

Typical Truck Loading Rack Installation
Foam-Water Deluge System with Diaphram Tank Stored Wet

Figure 3.20

Blessing and Curse

High expansion foam can be a blessing and a curse to firefighters. It can form a barrier against the radiant heat of a fire as firefighters advance toward the combustion, and firefighters can actually work surrounded in a blanket of high expansion foam because they can still breathe while completely covered by the foam.

The "curse" is that although people can still breathe in a high expansion foam blanket, vision is limited to almost zero. Firefighters (and occupants) can trip over unseen objects, fall in pits, or bump into objects.

High Expansion Foam Uses

High expansion foam has many uses and is an effective suppression and extinguishing agent on both ordinary combustibles such as paper and wood (Class A combustibles), and flammable liquids (Class B combustibles). This foam, (commonly referred to as *High X foam*), is most effective when used in enclosed or confined areas, because when used in an open area it will tend to flow out of the hazard area. If there is wind or a severe draft, it will blow away. However, high expansion foam is very effective fire protection against jet fuel spill fires in aircraft hangars, even with large hangar doors open. When used in aircraft hangars, the discharge of foam must be adequate to cover the hangar floor, build up at the rate of 3' per minute, and discharge for 10 minutes.

High X foam has proved to be efficient protection in roll paper storage warehouses as well. The foam discharge fills all the spaces between the rolls of paper and eventually covers the tops of the rolls. Sprinkler discharge would soak the paper rolls and ruin the paper stock. A more serious consequence would be the swelling of the paper rolls as they become water soaked. This occurred in a sprinklered roll paper warehouse of a newspaper company, and the paper rolls exposed to the heavy sprinkler discharge expanded and pushed out the concrete block walls. The consistency of High X foam does not soak the material with which it comes into contact. The exterior paper skin may be wet, but the roll is not soaked.

Sizing

To present the size of a high expansion system, a recent installation in a high grade rolled paper warehouse 100' long, 55' wide, and 50' high consists of:

- 4 high expansion foam generators. Each generator is capable of producing 13,500 cubic feet per minute of foam, for a total of 54,000 cubic feet per minute of foam.
- All generators are mounted on the roof.

This is one very important consideration with this installation, and every high expansion foam system discharging into an enclosed area. The air in the enclosure must be vented because the foam occupies so much volume. The described system is an excellent example—54,000 cubic feet per minute discharge of High X foam will displace 54,000 cubic feet per minute volume of air in the enclosure. The described installation consisted of eight good-sized roof vents to relieve the displaced volume of air.

Advantages

One advantage of high expansion foam is that it uses very little water and a conservative quantity of foam concentrate. The system consisting of the four large foam generators has a foam concentrate supply of 400 gallons, which is sufficient to fill the warehouse four times with high expansion foam.

Activation
High expansion foam systems are usually activated automatically by a signal from a detection system. Actuation may also be manual, and most installations have both automatic and manual activation.

When the high expansion foam system is activated automatically, the detection system should set off an evacuation alarm so personnel can at least commence exiting before the High X discharge begins. The detection system should also close doors, and when the discharge is in an enclosed area, a signal from the detection system opens the vents.

Air Supply
The foam generators can be either electrically or hydraulically operated, and it is imperative that the air supply to the generator be clean air, free of contamination from the combustion. If air supply to the generator could possibly be contaminated with fumes that would affect foam production, or products of combustion, it may be necessary to supply clean air to the generator through a duct.

Types of Foam
The properties and performance of a foam determine its selection for a specific application. There are numerous types of foam, and the following description of the most common foam types will help provide an understanding of their selection for a specific hazard situation.

Chemical Foams
Chemical foams are nearly obsolete, and are described for information purposes—the description will indicate why chemical foams have become obsolete.

To provide chemical foam, a chemical reaction takes place between chemicals like aluminum sulfate and sodium bicarbonate to produce the foam bubbles. The two chemical powders had to be kept in separate containers and injected into the water separately, producing two solutions that were supplied through two separate lines to the point of application where the two solutions were mixed to form foam. Proportioning the powder into the water at the chemical foam generator is dependent on the pressure of the water, and any back pressure in the system can critically affect foam formation.

The manpower required to operate chemical foam generators, the tremendous quantities of powders required to produce enough foam, and the fact that the foam generators had to be located close to the combustion so as not to have long lines producing back pressure at the generator all contributed to rendering chemical foam obsolete.

Mechanical Foam
Mechanical foam has almost completely replaced chemical foam because the formation of foam bubbles results from a mechanical means of introducing air into the foam/water solution. The mechanical means could be a sprinkler head, a foam nozzle, etc., where with chemical foam the energy to create the foam bubbles resulted from a chemical reaction.

Protein Foam
Protein foam is composed of naturally-occurring sources of protein—hoof and horn meal, feather meal, etc. The natural protein material is processed, and stabilizers and inhibitors added to resist bacterial decomposition, protect against freezing, and prevent corrosion.

Protein foams are not as easy flowing on a fuel surface as other low expansion foams, but protein foams do have excellent resistance against the foam blanket breaking up from radiant heat and reignition of the fuel.

Fluoroprotein Foam

This type of foam is similar to protein foam but has the addition of "fluorochemical surfactants" which produce an easier flowing protein foam. The addition of this surfactant not only allows the production of the air-excluding blanket of protein foam, but also produces a vapor-preventing film on the surface of the fuel.

Aqueous Film Forming Foam (AFFF)

Aqueous film forming foam (AFFF) is a synthetic foam concentrate. AFFF is a combination of fluorochemical and hydrocarbon surfactants, and no protein is involved. The AFFF is totally made up of synthetic products. Surfactants are chemicals that can alter the surface properties of water. Fluorochemical surfactants alter the property of water so that a thin film can spread over the surface of a hydrocarbon fuel even though the aqueous (water) film is more dense than the fuel.

Alcohol-Resistant Foam Concentrate

This is a foam concentrate used for fire suppression on water-soluble fuels that would drain off the water in the foam. Chemicals are added to the foam concentrate so that when the foam is applied to a water-soluble liquid, like methyl alcohol, a membrane is formed between the foam and the water soluble fuel.

Synthetic Detergent Foam

This is the foam concentrate used to produce high expansion foam. These foam concentrates do not produce foam blankets with vapor reducing films; they produce an aggregate of foam bubbles on the fuel surface.

CHAPTER FOUR

DRY AGENT SUPPRESSION SYSTEMS

CHAPTER FOUR

DRY AGENT SUPPRESSION SYSTEMS

In situations where water and other wet suppression systems are not practical methods of fire protection, dry agent suppression must be used. These include carbon dioxide (CO_2), dry chemical, and halon systems.

Carbon Dioxide Systems

There are many instances in which the use of an extinguishing agent can unnecessarily damage valuable equipment that is housed in the protected area. For example, an electrical equipment room, computer room, record vault, or storage room requires fire protection because of its operational value, the value of the equipment, and to prevent the spread of fire, yet water discharged on electrical or electronic equipment would considerably damage equipment not involved in the fire. In such cases, an agent such as carbon dioxide (CO_2) can be used to extinguish the combustion without causing unnecessary damage. When applied in the proper concentration, carbon dioxide will eliminate the combustion by displacing the oxygen required for combustion.

This section first presents an overview of CO_2 systems, followed by considerations, and then detailed methods of CO_2 system design.

Carbon dioxide gas is, in general, colorless and odorless. Impurities in some concentrations of dioxide and its chemical effect on the nostrils can produce the effect of an odor, but pure CO_2 is a completely odorless, inert, and electrically nonconductive gas which is 50% heavier than air.

Carbon dioxide has an impressive history of flammable liquid fire extinguishment because of its ability to form a temporary inert atmosphere above the surface of the liquid. If enough carbon dioxide is introduced into the electrical or computer room to displace enough of the room's normal atmosphere, the oxygen content of the room can be reduced below the 15% required for support of combustion and the combustion will cease due to oxygen starvation. The normal oxygen content of air is about 21%. Being a gas, carbon dioxide will not damage delicate, high value equipment or material, nor will it leave a residue.

Since there are no liquid or solid residues to clean up after the carbon dioxide discharge, subsequent damage and downtime after a fire is kept to a minimum. This advantage has led to the extensive use of CO_2 protection of food preparation areas such as deep-fat fryers, ranges, broilers, and cooking area exhaust ducts. In addition to its invisible characteristics, carbon dioxide is electrically nonconductive, which makes it an ideal fire protection agent for electrical equipment, particularly for oil cooled and oil insulated electrical equipment (such as oil switches, transformers and circuit breakers).

Carbon dioxide can be used to extinguish fires in all types of combustible materials, with the exception of chemicals and materials containing their own oxygen, which include cellulose nitrate and reactive metals such as sodium, potassium, magnesium and titanium.

Carbon dioxide gas is applied in one of two ways: total flooding or local application, depending on the physical characteristics of the room. Total flooding of an enclosed area with carbon dioxide provides three-dimensional fire protection. The carbon dioxide gas flows in and around all equipment and objects in the enclosed area. This means that equipment can be moved without relocating carbon dioxide nozzles or piping. However, there are instances where it is necessary to protect an isolated hazardous operation, a specific piece of equipment, or hazardous element of a total process, but where total flooding is not practical. Total flooding is difficult in areas that cannot be enclosed to contain the gas. The carbon dioxide system is designed to discharge gas *directly* onto the hazard to surround it with gas, to render the atmosphere inert in the immediate area, and extinguish the fire.

Local Application Systems

Examples of local application protection include dip tanks (see Figure 4.1) printing presses, motors, pumps, cooking ranges, and deep fat fryers.

The design of local application CO_2 systems is complex, requiring a detailed study, calculations, and design because the area does not remain inert for any appreciable length of time as is possible in a total flooding application. The CO_2 discharged in a local application system is soon dissipated, and its effectiveness therefore decreases rapidly. Detection of the fire and application of the CO_2 must be very rapid so that extinguishment takes place before the surrounding area absorbs sufficient heat to cause re-ignition after the CO_2 has dissipated.

Although in some cases a 30 second burst of CO_2 is suitable, it may be necessary to provide an extended discharge when re-ignition is a serious concern. For example, when there is a flammable liquid and surrounding masses of metal, an extended flow is necessary until the metal is cooled below the ignition temperature of the combustibles.

Other Considerations

The colorless gas quality of CO_2 is often questioned by those viewing a system discharge because it is usually accompanied by a characteristic thick white cloud. Therefore, it is of interest to understand the process by which CO_2 is released into the air. CO_2 is stored and transported through the system piping as a liquid. When this liquid reaches a nozzle, the pressure is released and the liquid carbon dioxide converts to a gaseous state. The ratio of expansion from liquid to gas is approximately 9:1. The white cloud and a fog-like vapor which may linger for a considerable time consist of solid particles of CO_2 "snow," which occurs when the moisture in the atmosphere has been condensed and solidified by the extremely low temperature of approximately $-10°F$ experienced when the CO_2 liquid is converted to gas at the nozzle. This low temperature will produce a frost on the CO_2 discharge nozzles, caused by condensation of the moisture in the surrounding atmosphere.

Figure 4.1

Low Pressure CO$_2$ System

Temperatures

CO_2 discharge temperatures are important to consider in system design. If the CO_2 nozzles are relatively close to metal parts that are in motion, the sudden and severe change in temperature could cause a distortion in the hot metal, causing the disintegration of the machinery and very possibly explosive disintegration in a machine such as a turbine generator.

The solid particles of CO_2 evaporate or change from solid to gas without passing through the liquid state very quickly, and in spite of the extremely low discharge temperature, the cooling effect of CO_2 does very little, if anything, in enhancing fire extinguishment. However, it may chill electronic equipment and disrupt its normal operation.

Grounding

Although it is not in a liquid state for long, discharge of liquid carbon dioxide can still produce an electrostatic charge, causing a spark. Therefore, when the CO_2 system is protecting an area that may contain a flammable liquid vapor (creating an explosive atmosphere), metal discharge nozzles should be used and the nozzles grounded. To dissipate any electrostatic charge resulting from the CO_2 discharge, all objects exposed to the discharge should be grounded.

Total Flooding Systems

The system using carbon dioxide to protect the electrical or computer room previously described is an example of *total flooding*. The entire room is filled, or total flooded, with a sufficient quantity of CO_2 gas to create an atmosphere which does not contain sufficient oxygen to support combustion. Total flooding of an enclosed area with carbon dioxide provides three-dimensional fire protection. The carbon dioxide flows in and around all equipment and objects in the enclosed area. This means that equipment can be moved without relocating CO_2 nozzles or piping. Total flooding systems can also be used for cabinet-type equipment such as a control board cabinet, computer equipment, and air ducts for cooking. In the design of total flooding carbon dioxide systems, 90% of the total enclosure must be enclosed in order to extinguish both flammable liquids and solids.

There are advantages and disadvantages to using a total flooding carbon dioxide system, all of which are discussed in the paragraphs that follow. However, it must be noted that each hazard situation is unique and a system should be chosen based on careful study of the location requirements, combined with a careful study of each hazard situation, a thorough knowledge of available extinguishing systems, and a system selection decision that is based on *the sum total of all conditions*, and not on the hazard only.

Hazards of Carbon Dioxide

Carbon dioxide activates the impulse to take a breath and inhale the oxygen the body requires. When the carbon dioxide concentration in our bodies reaches a certain level, it triggers the automatic impulse to breathe. It is this action that causes people to breathe rapidly in an atmosphere reaching a 3%–4% of carbon dioxide, and concentrations of approximately 9% can cause unconsciousness within a few minutes. At concentrations above 9%, unconsciousness occurs very quickly. In general, if the person is removed from the CO_2 atmosphere to a fresh air environment and given artificial respiration, recovery is usually quite rapid. However, a carbon dioxide concentration of about 20% will cause death in about 20–30 minutes.

Visibility is another hazard. If all of the people are not out of the area when the discharge occurs, the carbon dioxide snow and fog can interfere with visibility, and this combined with the rather frightening noise which accompanies the discharge, can alarm and disorient people trying to escape, easily leading to panic. Rapid breathing will ensue, accelerated by the CO_2, and unconsciousness will follow.

When an individual is suddenly exposed to relatively high concentrations of CO_2, it will produce mental confusion and loss of the ability to think clearly. This can have very serious consequences in a fire condition, and life safety protective measures must be given top priority in designing a total flooding carbon dioxide system.

Discharge Time Delay

To prevent injury to occupants, it is imperative to evacuate the area prior to the CO_2 discharge. This is accomplished during system design through the use of an audible and visual pre-discharge alarm. Detection actuates a pre-discharge alarm and a time delay device, and only after the time delay device has completed a predetermined time delay sequence does it activate the CO_2 discharge.

Several factors must be taken into consideration in determining an adequate, but not excessive, discharge time delay: the number of occupants that must be evacuated from the area, number of exits, travel distance to the exits, and the nature of the combustibles.

Another factor to consider is the spread of fire. The pre-discharge alarm feature, although absolutely necessary, can easily become a fire protection deterrent as the time can allow combustion to gain headway before the extinguishing action of the CO_2 begins. This becomes a major consideration when the material protected is in the Class A category, since deep-seated fires can result from combustion in paper, wood, and cotton. Because of its low cooling capacity, deep-seated combustion is difficult to extinguish with CO_2.

Re-ignition

The smothering action of carbon dioxide will quickly extinguish the flaming fire created by a burning flammable liquid, but to completely extinguish the smoldering or deep-seated fire may require maintaining the CO_2 concentration for a considerable period of time. Heat is still present to re-ignite the fire. For example, the flame stage when bulk paper is burning can be quickly extinguished with a CO_2 concentration as low as 0.125 pcf (pounds of CO_2 per cubic foot of area volume), but temperatures of 700°F to 1,000°F can still be present below the surface. As the CO_2 dissipates, the oxygen level builds up and re-ignition (or flashback) of the combustibles will occur. There have been cases where an individual can no longer see any sign of combustion and opens a door to enter the area. The heavier-than-air CO_2 rapidly flows out the open door and oxygen laden air rushes in to replace the CO_2, causing, in some instances, a combustion explosion.

Choosing the Time Delay

There are many variables to be considered in determining the length of time required for maintenance of the CO_2 concentration, and also the carbon dioxide concentration necessary to accomplish complete extinguishment in a deep-seated combustion condition. In most situations, the carbon dioxide concentration varies from 30% to 50% maintained from 30 to 60 minutes. There have been instances where an atmosphere close to a 100% CO_2 concentration had to be maintained for several days or weeks. Further criteria for choosing the time delay will be discussed in the carbon dioxide system criteria section.

Extended Discharge

When the CO_2 concentration and inert atmosphere are to be maintained for a relatively short time period, this may be accomplished by automatically introducing spurts of CO_2 into the area, or an automatically controlled continuous discharge. When the inerting period is prolonged over a long period of time, the CO_2 is usually introduced into the area manually, periodically procuring air samples to determine the oxygen concentration. These two methods are referred to as *extended discharge systems.*

An example of an extended discharge local application system is in the protection of a *turbine generator* used to produce electricity. These are massive machines that operate at a tremendous number of revolutions per minute, and the rotating shaft is lubricated by lubricating oil, piped and applied under considerable pressure. When a turbine generator is shut down, it continues to rotate for 30 to 45 minutes before coming to a complete stop, and during this run-down time, the shaft must be continually lubricated as the bearing-to-shaft tolerances are very close.

The majority of turbine generators are operated by steam power, which, coupled with the high operating speed, creates a tremendous heating of the metal, usually far in excess of the ignition temperature of the lubricating oil. A rupture in an oil line, or a loosened connection in an oil line fitting caused by vibration, will spew oil onto the hot metal, and ignition will usually follow. An atomized oil spray produces an extremely hot fire that is very difficult to extinguish. CO_2 applied immediately will smother the flame, and the turbine generator unit can be shut down immediately, but lubricating oil must be continually pumped to the shaft during the run-down period or the bearings will bind. As soon as the CO_2 atmosphere clears, the oil will re-ignite on the hot metal. Enough CO_2 must be furnished to provide an extended, continuous or intermittent, discharge to maintain an inert atmosphere during the entire run-down period.

In evaluating the protection of turbine generators or other rotating equipment, one should note that metal, when heated, expands, and conversely when cooled, contracts. Size and shape are distorted by radical temperature changes. This distortion can be controlled to perform useful functions, but uncontrolled it can be destructive, especially with moving metal parts operating in close tolerance, as with the turbine generator between the rotating shaft and the bearings. A rapid, uncontrolled, sudden cooling of the rotating metal shaft or the bearings could create enough distortion to bind the moving parts and cause serious damage to the equipment. Placing the CO_2 nozzles too close to heated parts in motion can initiate a rapid lowering of the temperature when the CO_2 is discharged, and for this reason some turbine-generator CO_2 local application protection systems are designed to vaporize the CO_2 liquid through an electric or steam vaporizer prior to its discharge on the bearings.

Factory Mutual Insurance Company recommends the use of water spray for protection of turbine generator bearings based on results of their research which provided no evidence of metal distortion. One of the major manufacturers of turbine generators states that water spray protection will definitely cause metal distortion and, therefore, it is not to be used as a fire protection system agent of their equipment.

This paradox is presented to accent the importance of treating each fire protection area carefully, basing system selection on the sum total of all hazard conditions.

Other Considerations

One design factor to keep in mind is that if the CO_2 protection is a total flooding system to protect a diesel or gasoline engine, the engine should be receiving its combustion air from outside of the room. If the CO_2 is sucked into the engine, there is the possibility of oxygen starved combustion and engine shut-down.

Warning signs are critical. Any room or area protected with a total flooding CO_2 system must be provided with warning signs to prevent personnel from entering the area following discharge, unless they are provided with self-contained breathing apparatus. Firefighters arriving at the scene of a total flooding CO_2 discharge must be made aware with obvious signs that they cannot enter the area without self-contained breathing apparatus.

Carbon Dioxide and System Criteria

The extinguishing capability of all carbon dioxide systems depends on the method of discharge into the fire area, the rate of application, and the total quantity of carbon dioxide discharged. The method of discharge is determined by the nozzles selected and their location; the rate of application is determined by engineering calculations that establish the pipe and nozzle sizing, and the total quantity of carbon dioxide discharged depends on the calculated demand and the storage capacities.

There are two methods of storing carbon dioxide which characterize the system as either a *high pressure* or *low pressure CO_2 system*. In general, choice of a system requires an evaluation of whether to use 20—75 pound high pressure cylinders requiring considerable storage space, and some very substantial maintenance for periodic weighing of the cylinders to determine their full capability, and maintenance time to replace the cylinders that have discharged their CO_2, or to use a small, 1,500 pound, low pressure tank.

High Pressure Systems

High pressure cylinders for CO_2 storage fixed pipe engineered systems are available in 50 pound, 75 pound, and 100 pound sizes. In considering the proper size of a cylinder to satisfy the required demand of a specific system, both the storage space available and the gross weight of the full cylinders are of importance. The use of 100 pound cylinders will reduce the required storage space, but because of the gross weight of the 100 pound cylinder, difficulty in handling becomes a factor. A 100 pound cylinder fully charged has a gross weight of about 300 pounds, while a 75 pound cylinder fully charged has a gross weight of about 215 pounds, and a 50 pound cylinder fully charged has a gross weight of about 165 pounds.

Carbon dioxide supply for a high pressure system is contained in high pressure cylinders which meet the test standards established by the ICC (Interstate Commerce Commission). See Figures 4.2 and 4.3. High pressure systems operate on pneumatic principles—pressure is used to open valves, release doors and dampers, and to operate switches and accessories. Operating pressures are multiplied by large pistons. Operating parts are located internally.

The liquid carbon dioxide is stored in the cylinders shown in Figure 4.2 at ambient or room temperature, and since ambient temperature determines the pressure in the cylinder, the container must be capable of withstanding the maximum pressure that could be expected. The recommended cylinder storage temperature range is 32°F to 120°F, with 130°F being the maximum permissible temperature. Each cylinder has a frangible disc as a safety valve with a bursting pressure of 3,000 psi, and at a maximum allowable temperature of 130°F, there is still a margin of safety below this 3,000 psi pressure point.

High Pressure Carbon Dioxide Cylinder Bank

(courtesy of Cardox Corporation)

Figure 4.2

There are two important factors to note when choosing high pressure tanks: 1) temperature and its effects on CO_2, and 2) tank capacity. Basically, the temperature of the CO_2 affects its state—liquid, gaseous solid, or vapor. In order to regulate the system and maintain the proper state for fire protection, the temperature must be controlled. In addition, temperature affects the volume of the liquid expansion in the tank. When filling high pressure CO_2 cylinders, a vapor space is necessary above the liquid CO_2 in the cylinder to allow for expansion of the liquid due to temperature changes.

The following explanation attempts to describe the basic effects of temperature on the state (and tank capacity) of the cylinders in high pressure systems. When the storage temperatures of high pressure CO_2 cylinders drop below 32°F, pressure decreases, and the rate of discharge can fall below design demand. If cylinder storage temperatures of below 32°F are anticipated, a method of compensating for the resulting pressure decrease must be provided by heating the area of cylinder storage. Cylinders must be maintained at a minimum of 32°F.

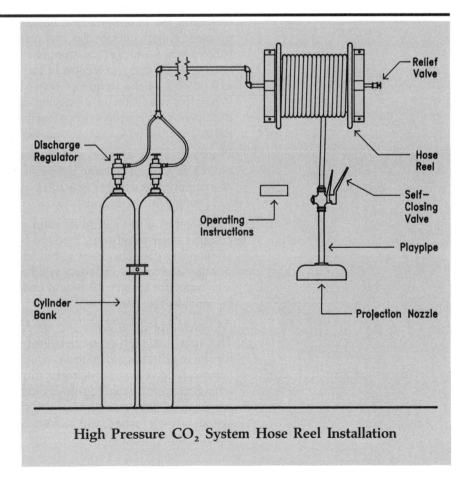

High Pressure CO_2 System Hose Reel Installation

Figure 4.3

What happens when temperatures increase? As the temperature of liquid carbon dioxide in the cylinder increases, the CO_2 vapor pressure increases. As the CO_2 vapor pressure increases, the density of the CO_2 vapor over the liquid CO_2 increases; the liquid CO_2 expands and its density decreases and turns to a gas. At 87.8°F, the *critical temperature*, the liquid and vapor have the same identical densities, and the liquid state has disappeared. Above 87.7°F, CO_2 is entirely a gas regardless of the pressure.

Pressure affects the state and temperature of CO_2. At pressures below 60 psi, CO_2 reaches a point called the *triple point*. When stored at pressures below 60 psi, carbon dioxide can be present in a solid, liquid, or vapor state. Below the triple point, carbon dioxide can only exist as a solid or a gas, depending on the temperature. At 60 psi, the liquid carbon dioxide that remains in the cylinder or container will convert to dry ice at a temperature of −69°F. If the pressure is reduced to atmospheric, about 15 psi, the temperature of the dry ice will drop to −110°F. When liquid carbon dioxide is discharged from the nozzles at atmospheric pressure, this same condition is established. When discharging CO_2 to extinguish a fire, the liquid converts to a vapor at approximately a 9: or 10:1 expansion ratio, and some of the vapor exists as dry ice particles at −110°F. The dry ice particles will sublime or change to gas without passing through the liquid state. The extremely low temperature created will also cause condensation of water vapor in the atmosphere, creating a fog which sometimes lingers long after the disappearance of the dry ice particles.

At 70°F, the pressure in a high pressure cylinder is 850 psi. Carbon dioxide expands as the temperature rises, but as long as an appropriate vapor space is maintained in the cylinder, the expansion does not increase the pressure. If the carbon dioxide cylinder was completely full of liquid CO_2, any rise in temperature, however slight, would rapidly increase the pressure in the cylinder. For this reason, the Department of Transportation (DOT) sets a limit on the amount of liquid CO_2 in a cylinder. The DOT criteria states that the weight of *carbon dioxide liquid* in a cylinder shall not exceed 68% of the weight of water that the container can hold at 60°F. While this is a factor in specifying design criteria, tank manufacturers label their pressure capacities according to how much CO_2 the cylinder holds, rather than the tank's actual volume capacity.

When a cylinder is listed as a 75 pound cylinder, it will contain 75 pounds of CO_2. The same is true of any cylinder capacity listing—if the description states that it is a 100 pound cylinder, there is a 100 pound CO_2 supply in the cylinder.

For example, if the hazard demand for CO_2 is 750 pounds of CO_2, the designer must select ten 75 pound high pressure cylinders. A reserve bank of cylinders may also be required (discussed later in this section), and the storage area for the cylinders will have to be large enough to accommodate twenty 75 pound cylinders, since the reserve supply must be equal to the primary supply.

Maintenance Considerations for High Pressure

The individual high pressure cylinders should be checked semi-annually for the required liquid content, which is done by weighing each cylinder. Most carbon dioxide equipment manufacturers provide a simple arrangement for weighing the cylinders without taking them out of service. The device consists of a very simple beam to which the individual cylinders are attached, and a direct reading scale. With a 10:1 ratio on the beam, a one pound pull would read 10 pounds.

When a cylinder weight has dropped by 10% of its required weight, it must be recharged, or even replace if the loss is due to a pin-hole leak. Other common causes of loss through leakage of carbon dioxide from cylinders are *corrosion* and *damage to the cylinders* (due to mishandling). This points out the necessity to secure the cylinders in an upright position to eliminate any possibility of their being knocked over, and selecting a safe storage area which must be outside the protected hazard area.

Cylinders are generally recharged at least at five-year intervals, and they should also be hydrostatically tested. After a system has been installed for 12 years without being activated, the cylinders must be discharged and each cylinder hydrostatically tested and refilled or replaced.

Reserve Cylinders

When a system with a demand for 750 pounds of carbon dioxide is activated by the detection system, manually or in the case of systems failure, the entire 10 cylinders will "dump." The hazard area is then without protection until the cylinders can be removed and replaced with charged cylinders. This time lapse without fire protection, in most cases, should be eliminated with the installation of a *reserve supply* of carbon dioxide cylinders equal in volume to the primary supply. (See the reserve supply shown in Figure 4.2.) The reserve cylinders also provide a second discharge of CO_2 should the primary cylinder discharge not have controlled the fire condition.

Switching from the main to the reserve supply is a simple manual operation—the reserve cylinders are similar to the primary cylinders in piping and installation. A toggle switch usually redirects the signal from the main detection system to the reserve cylinders. As with the main cylinders, a manual pull station will activate the reserve cylinders. There must be adequate visual indication that the remote station operates the reserve cylinder supply.

The cylinder manifold (common pipe to which the cylinders are connected) serves the main and reserve cylinders. The reserve cylinders are isolated with full flow check valves to prevent the charged reserve cylinders from discharging into the empty cylinders. As a further precaution, each manifold has a *bleeder valve* which will automatically close when the manifold pressure exceeds 20 psi during discharge.

Determination of the necessity for a reserve supply should follow a professional evaluation of the hazard relative to the frequency of fires, the area's need for uninterrupted operation, the availability of a human fire watch with hand extinguishing equipment and, most importantly, the speed with which the empty cylinders can be replaced with charged cylinders. In rare cases, it is possible to recommend eliminating the reserve supply, but this may be a "penny-wise-and-pound-foolish" decision, since fire ignition does not follow predictable and logical criteria.

How the System Works

To obtain an overview of how the system works, this section describes the process by which the CO_2 is propelled from the cylinder to the combustion. The pipes and fittings connecting the high pressure cylinders to the open nozzles are empty until the system is activated. When activated, the liquid carbon dioxide is transported to the nozzles through the piping. A siphon tube extends to the bottom of the cylinders so that liquid carbon dioxide will enter the piping. When this liquid CO_2 flows through the piping, there is a loss in pressure due to the increased volume of the empty piping and the friction between the moving liquid and the interior roughness of the pipe walls.

When the pressure of the liquid carbon dioxide drops, it changes to a vapor, and the piping fills with a mixture of liquid and vapor. As the mixture of liquid and vapor continues through the piping and the pressure continues to drop, the vapor increases in volume. This is the point where careful system design is critical. If the piping system is not designed properly, and the pressure reaches 75 psia (pounds per square inch absolute, which includes 15 psi atmosphere pressure) or 60 psi (the triple point) any further pressure loss would result in the remaining liquid flashing to a vapor and dry ice snow formation, which would clog not only the pipe but also the nozzle orifices.

Vaporization occurs as the first liquid carbon dioxide enters the piping and continues until an equilibrium is reached between the temperature of the pipe and the liquid carbon dioxide. The amount of liquid carbon dioxide vaporized, and the time required to reach this equilibrium with a resultant uniform flow, depend on:

- Flow rate
- Storage temperature
- Temperature, length, size, and piping material

It is necessary to adjust the total storage capacity and flow rate of the liquid carbon dioxide to compensate for vaporization and the time necessary to reach a flow equilibrium.

Design of the Piping System

The piping system must be designed to deliver the required rate of discharge through the selected nozzle orifices at the minimum nozzle pressure allowed, namely 300 psia for high pressure systems. This is the basic criteria for design of carbon dioxide piping systems (to be discussed further later in this chapter).

Once the required rate of discharge for the hazard is determined, and the piping and nozzles sketched, pipe sizes can be selected. The friction loss in fittings, valves, and cylinder connectors can be determined by changing the fitting or valve to an equivalent length of pipe. Tables are available to convert all sizes of fittings and valves to their equivalent pipe length, and charts will list the friction loss at various flows through various pipe sizes, with friction loss presented in psi per foot of pipe.

Terminal Pressures

Knowing the pressure at the CO_2 cylinders, and using the CO_2 manufacturer's flow charts for friction loss, the terminal pressure at the nozzles can be calculated. If the terminal pressure at the nozzle is below 300 psia, it will be necessary to revise the pipe sizes to compensate for friction loss.

The piping for a high pressure carbon dioxide system may be subjected to extreme pressures. Thus, in order to resist the maximum pressures that could be expected, the pipe and fittings must have a minimum bursting pressure of 5,000 psi. Valves that are under constant pressure must have a minimum bursting pressure of 6,000 psi. The piping system must also be corrosion-resistant and be able to withstand the temperature extremes of a CO_2 system.

Discharge Nozzles

Carbon dioxide discharge nozzles come in a wide variety of designs and discharge patterns. There are basically two classifications: *shielded, low velocity type*, and the *jet high velocity type*. Each discharge nozzle consists of the orifice and usually a configuration of a *horn, shield* or *baffle*. The design of a discharge nozzle depends on whether it will be used on a total flooding or local application system.

The manufacturer's nozzle orifice flow chart will indicate the proper nozzle orifice size required to deliver the desired flow rate at the nozzle pressure that was obtained from the piping calculations. *A note of caution*: If the manufacturer's orifice flow chart indicates that the nearest available nozzle size is larger than required, it will be necessary to recalculate the piping system to provide the specified minimum nozzle pressure.

Like the requirement for a piping system, carbon dioxide discharge nozzles should be of rugged construction to resist normal mechanical damage, extreme temperatures, and working pressures of the CO_2 system. The discharge orifice must be made of corrosion-resistant material. When nozzles are installed in areas where foreign material could clog the nozzle, they should be provided with frangible disc or blow-out caps that cover the nozzle orifice and are quickly ejected by the force of the discharging carbon dioxide.

Each discharge nozzle is permanently marked with the orifice size, and is indicated as the equivalent diameter. This equivalent diameter refers to the orifice diameter of the standard single orifice type nozzle which has the same flow rate as the nozzle in question. The standard orifice is an orifice with a rounded entry with a coefficient of discharge not less than 0.09.

Thumbnail Calculation

It is possible to produce a layman's estimate of the amount of carbon dioxide required for a total flooding system. Remember that this thumbnail calculation is for preliminary discussions only. The actual engineering study, design, and calculations require a seasoned professional's expertise.

To get a quick estimate of how much CO_2 will be required for a total flooding system (high or low pressure system), determine the cubic footage of the area and multiply this figure by 0.10 pounds per cubic foot of CO_2. For example: This thumbnail calculation does not take into consideration loss of CO_2 gas through enclosable openings and other factors that go into detailed calculations required for a CO_2 system. However, it will provide a preliminary demand of CO_2 for use in planning high pressure cylinder storage space, or the decision to use a low pressure storage tank instead.

High pressure CO_2 systems are more flexible than low pressure systems because selection of various sized cylinders — 50, 75, and 100 pound — makes it possible to provide almost the exact amount of CO_2 required to meet the system demand. Low pressure systems are addressed below.

Low Pressure CO$_2$ Systems

When the total carbon dioxide demand is so large that it is not economical to store in multiple high pressure cylinders, a low pressure system is recommended. Massive banks of high pressure cylinders not only take up valuable storage space, but also require considerable maintenance time and labor, testing cylinder weight and replacing cylinders when necessary.

Low pressure carbon dioxide storage is a single refrigerated storage unit that is available in several storage capacities, ranging from 500 pounds to 52 tons. These storage tanks have their own built-in refrigeration unit which maintains the liquid carbon dioxide at 0°F, and a pressure of approximately 300 psi. (See Figure 4.1.)

The low pressure liquid carbon dioxide storage tank is insulated, and refrigerant circulates through coils near the top of the tank. As the carbon dioxide vapor condenses at the coils, the tank pressure will vary. This tank pressure variance activates a pressure switch and automatically controls the compressor that circulates refrigerant through the coils.

The liquid carbon dioxide storage tank contains sufficient agent to supply the largest demand of a single hazard, or the individual demands of several hazards, plus a reserve supply of liquid CO_2. With the temperature controlled, filling density is not the serious consideration that high pressure cylinder storage presents. In addition, the low pressure unit has pressure relief valves to provide vaporization with subsequent cooling of the liquid, which prevents the storage tank from becoming liquid full. As a result, filling density for the low pressure storage unit is between 90% and 95%.

The Storage Tank and Refrigeration

The storage tank must be constructed, tested, equipped, and labeled in accordance with the Code for Unfired Pressure Vessels for Petroleum Liquids and Gases Specifications of the American Society of Mechanical Engineers (ASME). It must be equipped with a liquid level gauge, pressure gauge, and a high-low pressure supervisory alarm which activates an alarm if:

- the pressure in the tank rises to approximately 315 psi, or
- the pressure drops to 250 psi.

The refrigeration system is automatically controlled to maintain 0°F in the storage tank under conditions where the highest anticipated tank storage area temperature is attained. If the low pressure storage unit is installed in an area where the temperature is expected to drop below 0°F, an automatic heating unit will be installed, capable of maintaining 0°F in the pressure tank.

Should the refrigeration unit fail, and the pressure start to rise, relief valves on the unit are set to prevent the pressure from exceeding safe limits. When the pressure exceeds 341 psi, a diaphragm valve bleeds off the excess pressure, and if the pressure should continue to rise, a pop-off valve will operate at 357 psi, allowing a more rapid release of pressure. A final safeguard is a frangible disc which is designed to burst at 600 psi and vent the vapor to the atmosphere.

Where reliability of the refrigeration unit is a serious concern, dual refrigeration equipment is provided. The main concern for the reliability of refrigeration equipment is replacement of parts and down time required should the refrigeration unit need repair. Low pressure carbon dioxide storage tanks are so well insulated that even without the mechanical refrigeration, a four ton storage unit will not lose any carbon dioxide for at least 24 hours, and at an 80°F outside air temperature, this size unit will only lose approximately 100 pounds per day.

Pipes and Fittings

The pipe, fittings, and valves supplying carbon dioxide from the low pressure storage unit to the nozzles must have a minimum bursting pressure of 1,800 psi, compared with the minimum bursting pressure of 5,000 psi for pipe and fittings and 6,000 psi for valves in a high pressure carbon dioxide distribution system.

How the System Works

A low pressure carbon dioxide system is usually provided with a primary supply of carbon dioxide to protect the major hazard, several smaller hazards, hose stations, and a reserve supply of equal volume. When a single low pressure unit supplies multiple systems, this is accomplished through the use of selector valves.

The master selector valve located at the storage unit automatically directs the CO_2 to satellite selector valves located in the immediate vicinity of each individual system. The master selector valve is activated when a satellite selector valve is opened by a signal from the specific hazard detection system or by manual operation of an individual hazard system through an electric/manual pilot control system.

Dual Purpose CO_2 Units

An interesting use of the low pressure carbon dioxide system in a power generating facility is its dual use as a fire protection CO_2 supply and a CO_2 supply for use in purging the generators when they are emptied of the hydrogen atmosphere which is maintained internally when a generator is in operation. Hydrogen is commonly used as a cooling medium in large electric generators because of its very low wind resistance to the rotating equipment, and because it is a nonconductor of electric current.

Hydrogen is a highly flammable gas, and when the generator has to be worked on, it is imperative that all the hydrogen be evacuated. This can be accomplished by introducing CO_2 into the generator. The hydrogen is cleaned out or purged, and an inert atmosphere of CO_2 is created, nullifying the flammability of any hydrogen that might remain.

These dual purpose low pressure CO_2 units are sized so that the amount of CO_2 used for purging does not infringe on the supply of liquid CO_2 required for fire protection. A fire is possible in any number of protected hazards throughout the power generating facility, and especially during or immediately following the generator purging operation. By the use of shut off float valves, the supply of liquid carbon dioxide for purging cannot use any of the supply reserved for fire protection. This same arrangement is used when the tank contains both the primary supply and the reserve supply of CO_2.

To prevent any chilling effect or dry ice snow when the CO_2 is introduced into the generator during the purging operation, or when CO_2 nozzles protect hot bearings, the liquid CO_2 is passed through an electric or steam vaporizer which converts the liquid CO_2 to a vapor. This vapor is then forced into the generator or discharged on the bearings entirely as a CO_2 gas.

Inerting an Area with Carbon Dioxide

Carbon dioxide is sometimes used as a fire *prevention* agent. Spontaneous heating and combustion in bunkers storing coal can present a costly fire condition that is very difficult to control and extinguish. Applying large volumes of water is the only reliable method, and it is rather obvious what large quantities of water can do to a coal supply, especially if it is finely ground coal.

Natural spontaneous combustion can occur in coal storage bunkers due to leakage of air at the bottom entrance gate of the bunker or through the top of the bunker. By introducing CO_2 gas through a number of injection points, it is possible to have the CO_2 vapor spread throughout and completely fill the bunker, even a bunker filled with finely ground coal. By maintaining a low continuous rate of application of CO_2 vapor (normally 50 to 100 pounds of CO_2 per hour), an effective inert atmosphere (insufficient oxygen to support combustion) can be maintained in the bunker. The air leakage through the bottom entrance gate can be reduced or even eliminated by the injection of carbon dioxide vapor into the hopper just above the gate.

Carbon Dioxide Hose Stations

The extinguishing capability of small hand held CO_2 extinguishers is limited to small fires. The discharge range and duration of discharge of manual systems are quite limited—a few feet of range and a few seconds of discharge. The use of CO_2 hose stations will allow an operator to apply a large capacity of CO_2 from a safe distance away from the combustion.

Carbon dioxide hose stations can be supplied from high pressure CO_2 cylinders, or from a low pressure CO_2 tank. The hose reels are filled with CO_2 by an automatic or manual release. The manual release may be a pull station, push button, or similar device located adjacent to the hose reel and clearly identified by a permanent sign. (See Figure 4.3.)

One familiar automatic device is the bracket that holds the extended length of hose and discharge nozzle, commonly called the *discharge horn*. When the discharge horn is removed from the bracket, the CO_2 is automatically released into the hose. The discharge horn nozzle is equipped with a squeeze-grip type, quick opening and closing valve, so that the operator has complete control of the CO_2 discharge.

The supply of CO_2 for a hose station shall be sufficient to supply the discharge rate for at least one minute. If the hose station is to be used by an inexperienced person who might waste some CO_2, the supply should be increased. Multiple hose stations supplied from a single supply should have a supply adequate to accommodate all hose stations likely to be used at one time, at the determined discharge rate for one minute.

The hose is a high pressure type with a bursting pressure of 5,000 psi if the supply is from high pressure cylinders, and 1,800 minimum bursting pressure if the supply is from a low pressure tank. Three-quarter inch hose will permit a 100 to 150 pound-per-minute discharge rate of CO_2, and by using a 1" or larger hose, discharge rates of 250 to 300 pounds per minute with a discharge rate of 25'.

The location and number of hose stations should be determined by the equipment being protected. If possible, the hose reel should be placed close to an exit door so that if smoke and toxic fumes from the combustion become excessive, the person using the hose can quickly exit the area. Anyone using CO_2 hose stations should be properly trained in fire fighting and safety precautions. CO_2 is 50% heavier than air and will flow like water to low points in the building such as pits, basements, tunnels, etc. This caution must be seriously considered when there has been a considerable amount of CO_2 discharged from a local application system and/or CO_2 hose stations.

The Detection System and Air Leakage

The detection system that activates automatic CO_2 total flooding systems provides the alarm, but always specify a pressure switch on the discharge piping of the CO_2 system to activate the alarms. If someone manually activates the CO_2 system by a mechanical device, the detection/alarm system will be bypassed and in all probability the CO_2 will be discharged without activating any alarms.

The pressure switch to activate the alarm system can also be used to shut down the ventilation system in a room protected with a total flooding CO_2 system. It may also release magnetic door holders that hold self-closing doors in the open position, allowing them to close automatically. The magnetic door holder releases its magnetic hold on the door when a signal is received from the pressure switch.

An engineering evaluation of the protection may mandate that the HVAC (heating, ventilating, and air-conditioning) system be shut down, dampers closed, and magnetic door holders released as a result of a signal from the detection system, so that the room or area will be as tight as possible when the CO_2 discharges. Operating from the pressure switch activated by the discharge pressure of the CO_2 may present a situation where the doors close too slowly, HVAC fans keep rotating after power is shut off, and too much CO_2 will be lost, effecting the rapid build-up of the required CO_2 concentration.

CO_2 System Design

After clarifying the various types of CO_2 systems (low and high pressure, and total and local applications), we can now explain the details involved in designing these systems.

Local Application System Design

As previously described, local application carbon dioxide systems are used to protect against individual hazards, hazardous operations, processes, and equipment where it is impractical to enclose them and use a total flooding system. Total flooding may be impractical due to the low hazard conditions that exist in the surrounding area where the hazard is located, the number of people working in the area, the impossibility of shutting down machinery prior to leaving the area, or just the economics involved because of the size of the total area that would have to be total flooded with CO_2. However, the local application system does not have the advantage of walls that confine the CO_2 (as does the total flooding system), and the CO_2 discharge is rapidly dissipated. As a result, the design of local application hazard protection can be complex.

The local application system consists of CO_2 nozzles that discharge directly onto or into a specific hazard or hazard area. The hazard can be the surface of a flammable liquid or a piece of equipment. There are basically two methods of designing a local applications system: the *rate-by-area method* and the *rate-by-volume method*.

Rate-by-Area Method

When the hazard involves a flat surface or hazards basically oriented on a horizontal plane that can be readily measured, the rate-by-area method is used for design. Examples of such hazards are flammable liquid dip tanks and drain boards, quench tanks, open tanks, etc.

Before calculating and designing a local application CO_2 system by the rate-by-area method, it must be clearly understood that each type of CO_2 discharge has an approval rating which indicates its individual design data. Design data consists of:

- Rate of discharge
- Spacing limitations
- Maximum area of coverage at specified heights
- Design flow rate versus height or distance from the hazard surface
- Projection angle versus distance and height from hazard

Below are the steps involved in designing a local application system using the rate-by-area method.

1. **Extent of Hazard Area.** The first step in using the rate-by-area design method is to carefully determine the extent of the hazard area including not only the hazard itself, but also fringe areas that may be involved and any discharge obstructions that could be present in or adjacent to the hazard.

2. **Prepare a Layout.** The second step is to prepare a layout, drawn to scale, of the hazard and any ramifications of the hazard which would effect the discharge patterns of the nozzles. With this layout, nozzles can now be spotted around and/or above the hazard. Tankside or linear-type nozzles, if applicable, can be strategically located to take full advantage of their approval listings as to discharge rate, coverage, and discharge pattern.

3. **Area of Coverage.** The use of overhead nozzles located above the hazard are restricted by the height and area limitations stipulated in the approval listings for each nozzle. The approval listings furnish information relative to area of coverage at various heights. The area of coverage from overhead nozzles is considered to be in the form of a square.

4. **Height of Overhead Nozzles.** The height of overhead nozzles above the hazard determines the optimum flow rate for extinguishment of the specific hazard material being protected. This relationship of height, optimum flow rate, and hazard material is given in the individual approved criteria of each nozzle.

The general criteria for determining the placement of overhead nozzles for a rate-by-area design is as follows:

1. Distance between nozzles shall not be more than the square root of the area rating.
2. Distance to the edge of a hazard shall not be more than half the distance calculated in (1) above.
3. If possible, overhead nozzles should be installed over the center of the hazard and perpendicular to the hazard surface.
4. Overhead nozzles installed at an angle to the hazard shall not be installed at an angle less than 45° from the surface plane of the hazard.
5. The distance from the face of the overhead nozzle to the aiming point of the nozzle on the hazard surface shall be used to determine the height with reference to determining required flow rate and area coverage.
6. When installing overhead nozzles on an angle, use Figure 4.4 to determine the distance of the aiming point from the near side of the surface of the hazard by multiplying the width of the nozzle coverage by the appropriate aiming factor.

5. **Minimum Discharge Time.** The minimum discharge time for the nozzles is set at 30 seconds, but the hazard material may dictate an extension of this minimum. Metal heated by the fire before extinguishment may demand an extension of the discharge time in order to cool the metal below the ignition temperature of the hazard material. All forced drafts, unusual air movement, and wind currents in the area of the hazard must be carefully evaluated and compensated for by additional nozzles or relocation of nozzles to protect areas outside the hazard areas.

6. **Total Nozzle Flow Rate.** Once the discharge duration time is established, and the flow rates of the individual nozzles are determined, the total carbon dioxide supply required becomes as a simple multiplication problem of total nozzle flow rate demand times duration of discharge.

7. **Sizing and Routing of Piping.** The number of cylinders has now been determined, and a location for the cylinders remote from the hazard area is established. All that remains is sizing and routing of the piping from the cylinders to the nozzles. This phase of the project is the most complicated, and the part requiring the most accurate, exacting, and precise computations. Carbon dioxide equipment manufacturers prepare tables, charts, and graphs which are a help in these calculations, but the expertise of an engineer experienced in CO_2 systems is required in all cases except the simplest project.

Rate-by-Volume Method

The rate-by-volume design method of local application is used when (1) the hazard is three-dimensional or an irregular physical configuration that cannot be accurately measured and reduced to equivalent surface areas, or (2) hazards are partially enclosed but have openings that are too large to consider a total flooding system. Examples of such hazards are transformers, printing presses, and rolling mills.

The steps involved in designing a local application system using the rate-by-volume method follow.

Aiming Factors for Angular Placement of Nozzles, Based on 6" Freeboard	
Discharge Angles*	Aiming Factor**
45–60°	1/4
60–75°	1/4–3/8
75–90°	3/8–1/2
90° (perpendicular)	1/2 (center)

* Degrees from plane of hazard surface.
**Fractional amount of nozzle coverage area.

NFPA Carbon Dioxide Extinguishing Systems: Standard 12
(courtesy of National Fire Protection Association.)

Figure 4.4

1. **Assume an Enclosure.** The first step is to assume an enclosure around the hazard with the imaginary walls and ceiling a minimum of two feet from the sides to the top of the hazard. Now that the hazard has been boxed-in, compute the volume enclosed in the imaginary box. This volume calculation is the total volume of the assumed enclosure with no deduction for the volume of any solid object within the enclosure. If the hazard is small, a minimum four square foot enclosure shall be used.

2. **Set Minimum Discharge Times.** Once the cubic footage of the assumed enclosure is determined, the standard total discharge rate of carbon dioxide is one pound per minute per cubic foot of assumed volume for a minimum of 30 seconds. This discharge rate of one lb/min./cu. ft. is National Fire Protection Association criteria, which differs from the Factory Mutual Insurance required discharge rate of two lb/min./cu. ft. As explained in the rate-by-area design section, this minimum discharge time may have to be increased depending on the hazard conditions, forced drafts, or unusual air currents. The assumed volume shall be increased to compensate for the loss of CO_2.

3. **Permanent Walls.** If there are permanent walls enclosing part of the hazard, and if these walls extend at least 2' above the hazard, the total discharge rate of one pound of CO_2/minute/cu. ft. may be reduced proportionately, but never below 0.26 pounds/minute/cu. ft., and only after a careful evaluation of the hazard material and conditions. Any permanent wall which forms a partial enclosure must cover one entire side of the imaginary enclosure, and the length of this wall must be at least 25% of the perimeter of the enclosure. If one wall complies with this criteria, and there are adjoining walls that comply with the height requirement but do not extend the full length of the hazard area, these short walls may be included in calculating the percentage of the perimeter that is enclosed.

   Walls that meet the length and height criteria but are opposite each other and therefore not adjoining can only be considered as one wall. If the two opposite wall lengths added together equal at least 50% of the imaginary hazard enclosure perimeter, their total length may be used in calculating the required rate of discharge.

The discharge rates of these semi-enclosures for calculations based on the rate-by-volume method are calculated using prepared tables found in NFPA standards.

If the base of the semi-enclosed or imaginary hazard enclosure has openings that exceed 3% of the total base area, the rate-by-area method of calculations must be used to compute the discharge rate. If the openings in the base or floor are 3% or less of the total base or floor area, 10 pounds per minute per square foot of opening additional CO_2 must be included in the discharge rate calculations.

If a permanent wall forming a partial enclosure of the hazard has openings which total more than 10% of the wall area, the wall cannot be considered as part of the enclosure. If the total wall openings equal 10% or less of the total wall area, two pounds per minute per square foot of opening additional CO_2 must be included in the discharge rate calculations.

Special Considerations

Dip tanks and drainboards may have coated parts that hang on a conveyor above the flammable liquid surface. In this hazard consideration, the flammable liquid surface CO_2 protection will be figured using the rate-by-area method. If the coated parts extend more than 2' above the flammable liquid surface, this area must be protected by computing the discharge rate using the rate-by-volume method. An imaginary enclosure is again used, assuming the base to cover the same area as the protected surface and extending from 2' above this surface to the maximum height of the coated parts, and calculating two pounds of CO_2 per minute per cubic foot of the total volume of this assumed enclosure as a rate of discharge.

Hoods over kitchen ranges and hoods over special processes requiring exhaust systems (to collect and exhaust fumes so they do not permeate the room atmosphere) must be protected if the cooking facilities involve flammable grease or fat, or if the special processes involve flammable liquids.

If the area below the hood is protected with carbon dioxide (such as deep fat fryers, process burners, or liquid heaters), the rate of discharge in the hood is figured as 0.6 pounds of CO_2 per minute per cubic foot of hood volume. If the area below the hood is not protected, the rate-by-area method is used to calculate the discharge rate, using the entire interior surface of the hood.

Total Flooding System Design

Total flooding system design requires a thorough knowledge of the hazard, the enclosure, type of fire expected, enclosure leakage, and ventilation. Calculations can be quite accurate because the volume need not be assumed, but can be calculated, and the loss of carbon dioxide from the enclosure is a known factor.

The design and calculations for a total flooding system must be adequate to provide the required concentration within one minute of the start of the discharge into the room or area for surface fire conditions. For deep-seated fires, the design concentration shall be completed within seven minutes, with a rate of discharge to provide a 30% concentration within two minutes.

The temperature of the area to be protected with a total flooding CO_2 system will effect the fire suppression efficiency of carbon dioxide. To compensate for relatively high temperature areas, an additional 1% of carbon dioxide is required for every 5° over 200°F in the temperature of the area being protected.

Following are the steps involved in designing a total flooding system.

1. **Percent of CO_2 by Volume.** Charts published by the NFPA and carbon dioxide manufacturers indicate the various carbon dioxide concentrations, or percent of CO_2 by volume, required to extinguish common liquid and gas surface fires by total flooding. It will be noted that these charts indicate "Theoretical" and "Minimum Design" concentrations. The reason for this is the fact that some allowance is needed to compensate for the lack of complete and perfect diffusion of CO_2 throughout the entire area being total flooded at any point in time during the inerting period.

2. **Tight Enclosure.** As mentioned previously, to attach effective total flooding, the area must be a relatively tight enclosure. HVAC (heating, ventilation, and air-conditioning) systems must be shut down automatically before the CO_2 discharge commences. If this cannot be accomplished, additional carbon dioxide must be provided to compensate for the CO_2 lost through the HVAC system (due to HVAC fan run-down after being shut off, thereby causing a loss of CO_2).

3. **Stoppage of Flammable Materials.** In addition, flammable liquid pumps, conveyors, and mixers must also be stopped to discontinue the flow of combustible material into the area. Windows, dampers, and doors are shut automatically with pneumatic release devices actuated by CO_2 piping discharge pressure. The pressure in the CO_2 piping when discharge commences enters the pneumatic release device, and this pressure performs mechanical work which results in an electrical contact; electrical power is supplied to alarms, open dampers, and door closers.

4. **Rate of Discharge.** The total flooding rate of discharge for surface fires must be ample to reach the minimum discharge concentration within one minute after start of discharge. Flow equilibrium (liquid-to-vapor) in the piping with high pressure systems is usually an insignificant time element, and therefore the rate of discharge will be the total required volume divided by one minute.

$$\text{Rate of discharge} = \frac{\text{total required volume}}{\text{one minute}}$$

Low pressure systems can require a significant amount of time and CO_2 quantity in vaporization and cooling of the piping before a flow equilibrium is reached to demand a calculation to increase the equilibrium flow rate so that the recommended total quantity will be delivered within the one-minute-after-start of discharge requirement. In other words, delay time and weight of CO_2 vaporized during initial discharge must be included in the rate of discharge calculation.

5. **Time Delay of Vaporization.** Using the tables and formulas found in NFPA Standards and Factory Mutual Insurance data sheets, calculations will determine the time delay of vaporization in the piping and weight of CO_2 required to overcome the loss through vaporization in the piping.

6. **Concentration Level for Deep-Seated Fires.** In order to extinguish the deep-seated or smoldering fire, the total flooding system must provide a 30% concentration within two minutes from start of discharge, and this rate of discharge can be computed on the basis of 0.042 pounds/cu. feet. As mentioned previously in the discussion of total flooding systems, when the required design concentration is reached, surface fires will be extinguished, but with deep-seated or smoldering fires it is necessary to maintain the concentration for an extended period of time in order to "soak" the material with CO_2 until it reaches the deepest penetration of the combustion in the material. Thirty minutes is the average holding time for closely packed combustible material and occupancies such as fur vaults and record storage.

Calculations for the concentrations necessary to extinguish deep-seated combustion are difficult to establish, as the concentration required varies with the mass and physical arrangement of the material. Dense mass material has a high thermal insulating effect which permits combustion to retain the heat produced and prevents rapid soaking of the mass with the CO_2 gas. Tests have been conducted using the most common materials found in record storage vaults, dust collectors, fur storage vaults, cable rooms where insulated wiring is found, and small electric motors and machines, to establish the criteria to be used in designing total flooding systems for these hazards. As a result of these tests, it has been established that design concentrations of between 50% and 75% and flooding factors of between 0.083 and 0.166 are used in calculating the quantity of carbon dioxide required for each specific hazard listed above. Examples: electric equipment and wiring, 50% design concentration, 0.083 pounds of CO_2 per cubic foot of volume. For storage vaults, 75% design concentration, 0.166 pounds of CO_2 per cubic foot of volume.

7. **Quantity of CO_2.** To calculate the quantity of CO_2 required to extinguish a surface fire by total flooding in an enclosure, the following are the volume factors required for a few common materials.

Material	Minimum CO_2 Concentration(%)
Benzene	37
Ethel Ether	46
Gasoline	34
Kerosene	34
Lube Oil	34

Do not neglect openings in the enclosure that cannot be automatically closed at the start of discharge of the CO_2. The loss of CO_2 through these openings must be included in the total supply calculations at not less than one pound of carbon dioxide per square foot of openings.

HVAC systems that cannot be shut off automatically at the start of discharge of the CO_2, or cannot be immediately shut off manually, must be given careful consideration in computing the CO_2 supply. The volume moved by the operating ventilation, or HVAC, system during the time span between CO_2 initial discharge into the area and mechanical unit shut-down should be divided by the flooding factor, and this amount added to the total supply requirement. When the design concentration is greater than 34%, this calculated amount must be multiplied by a conversion factor found in NFPA standards.

8. **Unusual Temperature Conditions.** An additional design consideration is the existence of an unusual temperature condition in the enclosure to be flooded. For every 5° above 200°F the total calculated quantity of CO_2 must be increased by 1%. When the normal temperature of the enclosure is below 0°F, the total calculated quantity of CO_2 must be increased by 1% for every degree below 0°F.

General Design Notes

A few design notes to supplement the general discussion of pipe and fittings for high and low pressure systems.

General:

1. Pipe and fittings must be suitable for low-temperature use.
2. Pipe and fittings must be corrosion resistant inside and out.
3. Ferrous metals shall be galvanized.
4. Steel, copper, brass and metals with similar mechanical and physical properties are acceptable.
5. Copper tubing with approved flared or brazed connections is acceptable.
6. Malleable or ductile iron fittings are acceptable if they meet ASTM (American Society of Testing Material) A395, Grade 60-45-15.
7. Steel pipe that meets ASTM A-53 has excellent low-temperature characteristics.
8. Cast (gray) iron pipe and fittings are not acceptable.
9. Where pipe sections can be closed off, pressure-relief valves shall be installed to vent entrapped liquid carbon dioxide.
   a. High pressure systems: pressure relief set to operate at 2,400 to 3,000 psi.
   b. Low pressure systems: pressure relief set to operate at 450 psi.
   c. Care must be exercised to discharge CO_2 from pressure-relief valves without causing personal injury or mechanical damage.
10. The area required for venting of exceptionally tight enclosures can be calculated with the following formula. This formula is based on the assumption that expansion of carbon dioxide is nine cubic feet per pound:

    X equals Q divided by 1.3 times the square foot of P.

    X = Free venting area in square inches.
    Q = Calculated carbon dioxide flow rate in pounds per minute.
    P = Allowable strength of enclosure in pounds per square foot.

    Note: See Figure 4.5 for a guide to substituting a value for P. These values are based on general construction practice parameters.

This section has provided an overview of a CO_2 system—high vs. low pressure, and local vs. total flooding applications. This data can be used to generally make a choice between systems, however, in choosing the valves, pipes and design details, an appropriate engineer should be consulted.

Halon Systems

Halon, like carbon dioxide, is a fire extinguishing gas that is colorless, odorless, and leaves no residue, but there are several important differences between halon and carbon dioxide. Carbon dioxide extinguishes combustion by physically *suffocating* the fire: replacing the oxygen necessary for combustion with the carbon dioxide gas. Halon extinguishes combustion by *inhibiting* the chemical reaction of fuel and oxygen. The actual chemical reaction by which halon extinguishes combustion is not completely known, but this is one rather generic theory that is used in this text to describe the basic difference between the extinguishing action of carbon dioxide and halon.

Another important difference between halon and carbon dioxide is the effect of the gas on personnel, and this difference is one that has made halon popular as an extinguishing agent. By replacing sufficient oxygen to suffocate combustion, carbon dioxide also replaces sufficient oxygen to suffocate personnel. Since halon does not depend on oxygen replacement to extinguish a fire, the concentration of halon necessary to extinguish most surface fires is not harmful to personnel, an extremely important advantage, especially in total flooding systems.

Recall from the section on carbon dioxide systems that it is necessary to have a pre-discharge alarm with a total flooding carbon dioxide system. This was to provide time between the first alarm and discharge for personnel in the area to evacuate because once the discharge commences, the occupants were in danger. With Halon 1301, the most common and popular halon used today, the total flooding discharge concentration for areas such as a computer room or control room is not hazardous to the occupants.

There is a restriction on halon use, however, that affects its continued popularity. Halon 1301 is a member of a family of halogenated agents that has a detrimental effect on the earth's ozone layer, Halon 1301 production has been frozen by international agreement. In a few years the use of Halon 1301 will be prohibited completely.

At the present time, halon system use is limited, and these systems can now only be used if no other agent can provide protection. Most government agencies and environmentally conscious corporations completely prohibit the installation of new halon systems. Accordingly, when a Halon 1301 system is authorized, *testing* of the total flooding system becomes a major concern.

The testing of total flooding Halon 1301 systems now consists of pressurizing the enclosure to determine the concentration that will be retained for the required time period. This test does not verify that the concentration is within the safe 5% to 7% range, nor does it confirm that the concentration distribution throughout the enclosure is constant.

Use of Halon 1301

Concentrations of Halon 1301 of between 5% and 7% by volume are levels that can safely be inhaled by personnel for four to five minutes without serious cardiac or central nervous system effects. This means that

Strength and Allowable Pressures for Average Enclosures			
Type Construction	Windage	Pressure	psi
Light Building	100 mph	25 lb./sq. ft.*	.175
Normal Building	140 mph	50 lb./sq. ft.**	.35
Vault Building	200 mph	100 lb./sq. ft.	.70

* Venting sash remains closed.
**Venting sash designed to open freely.

Figure 4.5

NFPA Carbon Dioxide Extinguishing System: Standard 12

High Pressure System: *Pipe and Fittings*
1. Minimum bursting pressure 5,000 psi.
2. Steel pipe—Schedule 40 standard weight acceptable in size up to 3/4".
3. Schedule 80 extra heavy pipe for pipe sizes above 3/4".
4. Fittings in sizes up to 3/4" acceptable in standard 150-pound malleable-iron-banded.
5. One-inch and two-inch sizes: Extra-heavy, 300 pound, malleable-iron-banded.
6. One-half-inch to 2" sizes may be 300 pound ductile-iron-banded.
7. All larger sizes should be forged steel.
8. Valves constantly under pressure shall have a minimum bursting pressure of 6,000 psi.
9. Valves not under constant pressure shall have a minimum bursting pressure of 5,000 psi.

Low Pressure System: *Pipe and Fittings*
1. Minimum bursting pressure 1,800 psi.
2. Steel pipe continuously under pressure (between tank and master valve or master selector valve) shall be Schedule 80 with forged steel or welded fittings.
3. Piping between master and selector valves—Schedule 80 with 300 pound malleable or 300 pound ductile-iron-banded screwed fittings or Schedule 40 with standard weight welded fittings.
4. Open-end pipe (not under constant pressure)—Schedule 40 with 300 pound malleable or 300 pound ductile-iron-banded screwed fittings.
5. Between storage tank and selector valves, black steel pipe is acceptable.
6. Valves shall be capable of withstanding a hydrostatic test to 1,800 psi without permanent distortion.
7. Low pressure storage units are available from 500 pounds to 125 tons of carbon dioxide.
   a. Low pressure storage tanks can be filled from a transport truck with no interruption to protection provided by the unit.
8. If only one hazard is to be protected by a low pressure unit, and a direct line is run from the tank header to the hazard only a master selector valve is required, with normally de-energized pilot control. When more than one hazard is supplied from the same low pressure tank header, the master valve has a normally energized pilot control so it will open if electric power should fail and allow carbon dioxide to flow to the selector valves that control flow to each individual hazard. The selector electro-manual valves have de-energized pilot controls.
9. The master valve is designed to permit the liquid carbon dioxide downstream of the master valve to flow back to the tank when electric power is restored and the master valve closes, or following operation of the system. This automatic flow-back arrangement eliminates the need for relief valves in the piping.
   Note: The description of low pressure systems and low pressure equipment is based on Cardox Division of Chemetron Corporation, because at this writing this firm is the only manufacturer of low pressure CO_2 tanks and low pressure equipment.

(courtesy of National Fire Protection Association.)

Figure 4.6

discharge can commence immediately upon detection with no pre-discharge alarm time delay. Concentrations above 7% can cause slight dizziness or a feeling of being mildly intoxicated. When the concentration passes 10%, mental and physical impairment can occur. Concentrations above 17% are assumed to cause unconsciousness and possible death.

When it is necessary to provide protection in an area where personnel cannot cease activity or shut down equipment and evacuate the area immediately upon hearing the fire alarm, Halon 1301 discharge can commence combustion suppression when manually activated. As a result, total flooding Halon 1301 systems have found extensive use in areas like airport control towers and air traffic control centers, where personnel cannot jump up and run for the exit when the fire alarm is activated.

Need for Precise Design

When Halon 1301 is used as a total flooding system in areas where personnel are present and the concentration must not exceed 7%, it is absolutely necessary that the system be designed and calculated with a high degree of professional precision. Further, the design of a total flooding Halon 1301 system must be performed by an experienced professional because the flow calculations to provide the proper discharge at each nozzle with all but the simplest systems is extremely complex and sophisticated. These are so complex that they cannot be done manually, and must be performed by computer programs (which are checked by national testing laboratories to assure not only their accuracy, but also their limitations).

The purpose of this section is to acquaint the reader with the preliminary specification and design of a system. In no way should a lay person attempt to go beyond preliminary design. Figure 4.7 shows the complex assembly involved in halon system design.

Alarm Systems

Halon 1301 total flooding systems are automatically activated by a detection system, and in order to prevent false discharge, the detection system is usually designed to provide all alarms, automatically close self-closing doors to the enclosure by releasing magnetic door holders, and shut down heating, ventilating, and air-conditioning (HVAC) when the first detector is activated. Activation of a second detector releases the discharge.

The detection system provides the alarms, but since there is provision for manual operation of the discharge, it is best to always include in the specifications and cost estimate a pressure switch on the halon discharge piping. This switch will activate all the alarms and control devices that the detection system handles if the detection system is bypassed by manual operation.

In lieu of magnetic door holders, *self-closing doors* may be held open and automatically closed by using pressure trips that are connected by piping to the discharge piping of the system. When the system operates, pressure in the piping activates the pressure trip devices and the door closes. (These pressure trips can also be used with CO_2 systems.)

Whether the alarms are bells, horns, or sirens, it is important to have a visual signal, such as a strobe light, to warn personnel who might not hear the alarm or who have a hearing impairment.

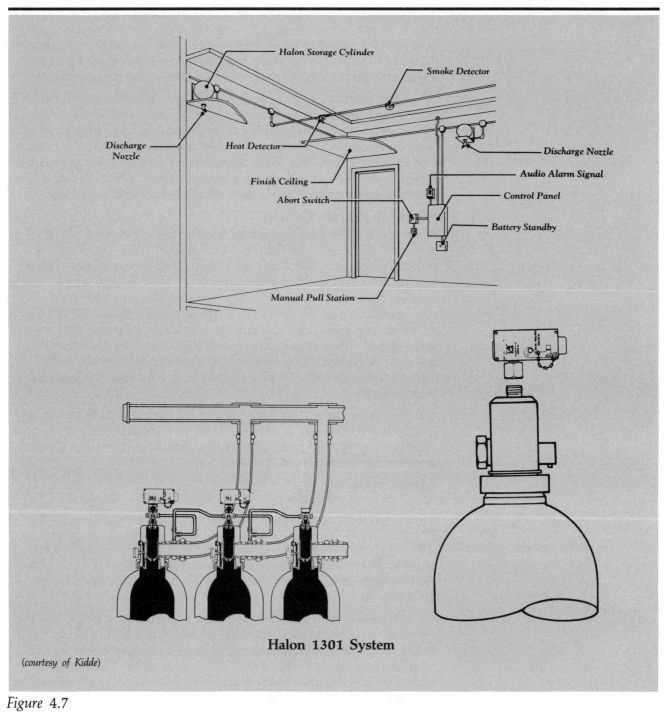

Halon 1301 System

(courtesy of Kidde)

Figure 4.7

136

An audible and visual alarm are also necessary immediately *outside* of the protected area to provide individuals outside the room with notification that the halon has discharged. The sign/alarm should post notice *not* to open doors and to *rescue* personnel who may need assistance in leaving the area. Responding firefighters also need this notification so they can use self-contained breathing equipment before entering the area.

Rapid Discharge Rate

The total flooding system shall be specified to establish the required concentration in the enclosed area within 10 seconds, and this concentration is to be maintained for a period of 10 minutes. One reason for this rapid discharge rate is the fact that Halon 1301 will decompose when exposed to temperatures in the 900°F range, and the decomposition products not only have an extremely irritating odor, but can also be very toxic. By developing the concentration very rapidly, the combustion can usually be suppressed and extinguished before materials exposed to the fire can reach high temperatures. Flaming combustion can be quickly suppressed, thereby reducing the heat of the fire.

The extremely rapid rate of discharge of a total flooding Halon 1301 system presents a condition that must be given careful consideration. When a Halon 1301 total flooding system is used to protect areas where the velocity of the discharge could damage delicate material, a very careful evaluation of the location of the nozzles, the discharge rate, and the type of nozzles must receive top priority.

For example, such a detailed evaluation was not developed for the recent installation of a Halon 1301 total flooding system in an archives storage room of an art museum. When the discharge test was conducted, the velocity of the discharge blew irreplaceable artifacts off the shelves and damaged them beyond repair. The extremely rapid rate of discharge of a total flooding Halon 1301 system must be given careful consideration.

The velocity of the discharge has been known to blow ceiling tiles off a suspended ceiling support, and the noise of the discharge can be extremely frightening, if unexpected. To counter this, there is customarily one type of alarm when the first detector is activated, and a different type of alarm when the halon is discharged, although the alarm in the protected area is quickly drowned out by the noise from the discharge.

Storage of Halon

Halon 1303 is stored in cylinders as a liquid and, to maintain the Halon 1301 in a liquid state and to provide sufficient pressure to transport the liquid Halon 1301 through the piping to the discharge nozzles. The pressurization is maintained at 360 psi at 70°F or 600 psi at 70°F. The 360 psi level is usually specified for systems with relatively short piping designs, and the 600 psi system is primarily used for complex piping designs or where there is considerable distance between the storage cylinders and the hazard.

If it is not possible to replenish the supply of Halon 1301 cylinders within a reasonable time (usually 24 hours) or if the hazard is extremely valuable to the operation of the facility, or the hazard has the potential for recurring fires, it is necessary to furnish reserve supply cylinders permanently piped into the system. The operation of a switch will change the system supply from the empty primary cylinders to the reserve cylinders. A check valve in each supply from the primary and reserve cylinders to the system piping will prevent the charge cylinders from discharging into the empty cylinders.

Control Switches for Storage

Manual pull stations are provided to trigger the system manually, and activation of a pull station transmits a signal to the cylinder discharge devices. Activation of a manual pull station will cause immediate discharge of the halon, and at the same time set off all alarms and control devices.

Each system has a control panel that receives signals from the detection system and transmits signals to the cylinder discharge devices, alarm system, and control systems.

Abort switches may be requested by an owner to abort, or put on hold, the discharge of the halon after a detector has activated the alarms. Automatic time delays to delay the discharge of the halon following activation of the detection system are acceptable, but the use of a device such as an abort switch to prevent halon discharge until a fire situation is confirmed is not acceptable, and usually not permitted by the insurance carrier.

When an abort switch is installed, it should be a "dead man" switch. A dead man switch must be held in the operating position by an individual, and once released it will automatically activate the halon discharge. If an individual hears the alarm when the first detector is activated and does not see any smoke or fire condition, he or she can hit the abort switch and prevent the halon discharge.

A note of caution: If an individual hits the abort switch, the only way they can release the switch without discharging the halon is to reset the panel that controls the system. If they are alone in the area, and it is obvious that a non-fire condition activated the detector, the abort switch should be within reach of a telephone or the control panel.

Never use a "positive" type abort switch similar to a toggle switch. There is too much probability that the switch would be left in the "off" position rendering the system inoperative.

Placement of Halon Cylinders

Halon 1301 cylinders should be stored outside the hazard area and secured with brackets designed to hold the cylinders. The storage cylinders should be safe from mechanical and physical damage, and placed in an area where the temperature will not exceed 130°F nor fall below − 20°F.

Storage Configurations

Systems with the halon cylinders located outside the hazard area, with the extinguishing agent piped to the discharge nozzles, are referred to as *engineered systems*.

In other Halon 1301 systems, the individual cylinders are installed in the hazard area. These containers may be either cylindrical or spherical in shape, and can be installed on the wall, above the suspended ceiling, or below the raised floor. Each cylinder is a self-contained "system" with supply, cylinder actuator, and discharge nozzle. The requirement for piping from the cylinder to the discharge nozzle may consist of a short nipple, or the nozzle may be part of the cylinder itself with no piping involved. Systems with the supply cylinders located in the hazard area are commonly referred to as *modular systems*.

Estimating Approximate Amounts

During the design of an engineered Halon 1301 system, the questions most frequently asked by the architect or the mechanical engineer are: How many Halon 1301 cylinders? How big are the cylinders (to allocate storage space)?

Below is a simple method by which to estimate the approximate amount of Halon 1301 required to protect the hazard with a 5% to 7% concentration:

1. Calculate the *cubic footage* of the enclosure. If there is a raised floor, as found in the majority of computer rooms, use the height of the area from the bottom of the suspended ceiling to the floor below the raised floor. The area above the suspended ceiling does not have to be included in the calculation for total cubic footage unless there are combustible materials above the suspended ceiling.
2. Once the cubic footage has been determined, multiply this figure by 0.025, and the answer will be the estimated amount of Halon 1301 in pounds.
3. Consult a halon manufacturer's catalog for the cylinder capacities in pounds, and once the appropriate cylinder is selected, the size will be indicated.

If the poundage of Halon 1301 is considerable, it will be necessary to divide up the total pounds required among several cylinders, if reserve supply is required an equal number of equal size cylinders will occupy the storage space.

This formula for estimating the required number of pounds of Halon 1301 includes about a 20% safety factor to account for normally encountered loss of halon concentration through various unsealable openings. This type of unsealable opening occurs below doors: exhaust fans that are shut down when the detection system is activated may continue to rotate, or ducts that have closed dampers during discharge of the halon often have a certain amount of leakage.

Loss of gas through these small openings is accented by the tremendous rate of discharge, so this 20% safety factor is necessary in figuring the approximate number of pounds of Halon 1301 required. Large openings that cannot be closed prior to the discharge must be carefully considered in the design calculations.

In areas with underfloor spaces, do not attempt to protect the underfloor space with above floor system. The underfloor space in usually used as ventilation for the machines on the raised floor, and a considerable amount of halon discharged in the area above the floor will spill into the underfloor space, but some of this halon will be sucked up into the machines and the below floor concentration will be diluted.

If the underfloor space is to be protected, it is best to install both detection and discharge nozzles at the ceiling for the above floor area, and detection and discharge nozzles below the raised floor. Activation of the detection either above the floor at the ceiling or below the raised floor must activate the entire system. The discharge nozzles at the ceiling for the area above the raised floor, and discharge nozzles below the raised floor, must activate simultaneously.

In requesting bids or detailed system installation design and calculations, do not neglect any openings in the area that cannot be dampered, shut down, or closed. This includes open drains and cable shifts. Also, be absolutely certain that raised floor spaces do not extend beyond the above floor partitions, as loss of halon into adjacent unprotected areas or areas beyond the scope of the system will dilute the concentration below that required to extinguish the combustion.

Partial Total Flooding Halon 1301 Systems

Halon 1301 is an expensive agent, and when the equipment to be protected is in a room of considerable height, for example 40', and the equipment height is only 15', the "total flood" of the room can be calculated to cover the equipment plus 5' to 10' (above).

Since Halon 1301 is considerably heavier than air, it will settle to the lower areas of a room. To ensure the accuracy of the calculations, design, and placement of the nozzles, experienced Halon 1301 design professionals must engineer the system.

An example of partial total flooding is the protection of aircraft flight simulators. The area containing these simulators is considerably higher than the simulators. The Halon 1301 is discharged to provide a 6% concentration to protect the sophisticated electronic equipment associated with the simulators, to a height of approximately 10' (above the simulators).

Research on Replacing Halon

The E.I. duPont de Nemours Company, who first developed Halon 1301 under a program sponsored by the U.S. Air Force to develop an agent for extinguishment of aircraft engine fires with the aircraft in flight, as well as other manufacturers, is working on a substitute for Halon 1301. This is a very difficult task since Halon 1301 has so many advantages. The cost of retrofitting existing Halon 1301 total flooding systems is another factor to consider.

Dry Chemical Systems

Dry chemical is a generic term because it includes several different types of dry chemical powders in extremely small particles. Dry chemicals have been used in hand extinguishers since the 1920's, but their use in fixed pipe extinguishing systems is relatively recent. The original and still most well known dry chemical is *sodium bicarbonate*, but there are other dry chemicals that have gained tremendous public acceptance. *Potassium bicarbonate*, also known as *Purple K*, is used extensively on hydrocarbon fires because it is almost three times as effective as sodium bicarbonate. *Monoammonium phosphate* is an all purpose agent, and is used in the majority of home fire extinguishers.

Classification of Dry Chemicals

Before proceeding with a discussion of dry chemicals and dry chemical systems, it is necessary to explain the lettering system used to identify different classes of combustibles:

> **Class A** combustibles are ordinary combustibles such as wood, paper, cloth, rubber, and some plastics.
> **Class B** combustibles are flammable liquids, paints, grease, solvents, etc.
> **Class C** is not combustible, but indicates that the agent is electrically nonconductive, and can therefore be safely discharged on energized electrical equipment, devices, and wiring.
> **Class D** refers to combustible metals such as magnesium, sodium, and potassium. Agents used for extinguishment of Class D combustibles are *not* referred to as a dry chemicals, but as *dry powders*.

Sodium bicarbonate is basically a Class B extinguishing agent, but it can handle small Class A fires that are on the surface of the combustible. Purple K is primarily a Class B extinguishing agent because of its efficiency in fighting flammable liquid fuel spill fires. Monoammonium phosphate is a multi-purpose agent because it is an agent used to extinguish Class A and Class B combustibles, and is safe when used on energized electrical equipment, so it is an A/B/C agent.

Dry chemical systems can be either total flooding or local application type.

The design criteria for a dry chemical total flood system closely follows the requirements for a carbon dioxide total flood system: 1) discharge of the agent into an enclosed space, 2) system activation automatically by a detection system with manual activation potential, 3) doors, windows, and dampers automatically closed, and 4) HVAC (heating, ventilating, air-conditioning) system shut down when system is activated.

Here the similarity ceases with a dry chemical total flooding system. With halon and carbon dioxide total flood systems, the required concentration in the enclosure must be maintained for a specified time period to prevent re-ignition of the combustion. Dry chemical concentration must be obtained in all sections of the enclosure within 30 seconds, but there are no requirements for maintaining the concentration because the dry chemical settles quickly. For this reason, when designing a dry chemical system sources of reignition must be evaluated along with the hazard itself and eliminated if at all possible.

The total flooding concentration for various hazards is mandated for Halon and carbon dioxide, but this is not possible with dry chemicals. The quantity of dry chemical, the rate of flow in the system, and the piping configuration for nozzle positioning are all dependent on the specific dry chemical used and the hazard being protected.

This text will provide the basic criteria for a total flooding dry chemical system, but experienced professionals must be consulted once the preliminary plans have been established.

Safety to Personnel

Safety to personnel is a high priority concern with dry chemical total flooding systems. The discharge of dry chemical systems can cause loss of visibility, and produce difficulty in breathing, resulting in a panic condition.

If personnel are present in the area protected with a dry chemical total flooding system, a pre-discharge alarm should be provided to allow evacuation before discharge.

Dry Chemical System Design

Most fixed pipe dry chemical total flooding systems, called *engineered systems*, are two cylinder systems: one cylinder contains the dry chemical, and the other contains either nitrogen or carbon dioxide.

When the detection system is activated, or the system is tripped manually, the nitrogen or carbon dioxide is automatically discharged into the dry chemical cylinder, pressurizing the dry chemical which is then carried through the piping to the discharge nozzles. However, this is not quite as simple an operation as it sounds. Before the dry chemical can be transported through the piping in a free flowing manner, the gas/dry chemical mixture in the cylinder must be in a fluid state. In this condition, the mixture is said to be *fluidized*. The dry chemical must not be fluidized before it can be released to the piping, but also the pressure in the dry chemical cylinder must build up to the design pressure required to transport the fluidized mixture through the piping, arriving at the nozzles at the required pressure.

To equalize the pressure in the dry chemical cylinder, a *pressure regulator* is usually provided to control the pressure of the nitrogen or carbon dioxide into the dry chemical cylinder. To accomplish the equalization of the pressure in the dry chemical cylinder, and to assure the fluidization of the dry chemical in the cylinder, the dry chemical cylinder has a bursting disc which seals it off from the piping until the calculated pressure is reached. When the predetermined pressure is attained in the cylinder, and the dry chemical has been fluidized, the disc will rupture and the dry chemical/gas mixture will enter the piping and flow to the discharge nozzles. The bursting disc (between the empty piping and the dry chemical cylinder) also prevents entry of moist air, condensation, and caking of the dry chemical.

The fluidized mixture is transported through the empty piping at extremely high velocities, upwards of 125 feet per second, at pressures of up to 350 psi. This high velocity and pressure, combined with the fact that although the dry chemical is technically a powder (in the gas/agent mixture) when in a solid state, produces a tremendous thrust and force at elbows and tees. The dry chemical system must be designed and installed with the necessary strength to contain this force.

The fact that the dry chemical is a solid makes it rather easy to visualize how the dry chemical powder by centrifugal force separates from the gas at an elbow and follows the curve. With the dry chemical and gas mixture flowing through the piping at tremendous velocity, and after being thrown against the outside curvature of the elbow (if the next fitting is a tee), the dry chemical may rush past the tee outlet hugging the outside of the straight part of the tee. To give the dry chemical a chance to re-mix with the gas before it reaches the tee, a straight piece of piping, not less than 20 times the pipe diameter in length, must be installed between the elbow and the tee.

If space limitations will not allow this length of pipe, a device called a *venturi* assembly is installed between the elbow and the tee. This venturi is basically a restricted "throat" which allows the dry chemical to re-mix before entering the tee.

With tees on a straight run of pipe, the dry chemical traveling at a high velocity in the straight pipe can pass over the outlet of the tee. To overcome this effect, tees are most effective when they are installed "bullheaded"—discharging into the side of the outlet of the tee.

Tests

The optimum amount of dry chemical and the rate of discharge for specific hazard have been determined by experiment and test for total flooding systems. The quantity of dry chemical, rate of flow, discharge rate, and nozzle placement resulting from these experimental tests are plotted on graphs.

The tests that provide information for these graphs are based on using different flammable liquid surface fire areas, the amount of dry chemical required to extinguish the combustion, and the minimum rate of discharge necessary to accomplish the extinguishment. Each manufacturer must conduct tests since every piece of equipment—nozzles, devices, and dry chemical compounds—affect the flow rate discharge capability for their specific system. The graphs developed from these tests can be obtained from the various dry chemical manufacturers for use in specifying and preliminary design of dry chemical total flooding systems.

The following formula was developed from the tests conducted by one manufacturer, and is offered here to provide the generic concept of testing for total flooding dry chemical systems. It is imperative that this formula be considered as an example only, and should not be used for all situations, all types of dry chemicals, or all manufacturers' systems: *The weight of dry chemical in pounds* for a total flooding system can be determined by *multiplying the cubic feet of the enclosure by 0.0385.*

As with all dry chemical total flooding systems, this is the gross cubic feet of the enclosure less the volume of permanent equipment or other permanent fixtures that occupy a considerable volume in the enclosure: This formula is presented for example only and should not be used for any preliminary calculations that will become part of a specification or cost estimate. Each specific total flooding dry chemical system calculation must be based on the manufacturer's formulas and graphs for the particular hazard and enclosure.

Local Application Dry Chemical Systems

Local application systems are used when there is no enclosure around the hazard or when the hazard is located in an area enclosure where total flooding would be impractical.

As with other local application systems, the dry chemical discharges directly onto the hazard. Dry chemical local application systems are used extensively to protect solvent and paint dip tanks, open tanks of flammable liquids, gasoline, and flammable liquid racks. Dry chemical local application systems have also been used to protect oil insulated electric transformers located in transformer vaults and exteriors transformers. When the system protects a hazard located outside, additional dry chemicals must be included to compensate for loss due to wind currents carrying some of the discharge away from the hazard.

As with total flooding systems, the design of local application systems depends on the type of dry chemical, the rate that the dry chemical is discharged, the nozzle type used, and the velocity of the dry chemical. All of these design and calculation factors, including the amount of dry chemical required for the hazard, are determined by experimental testing by manufacturers and testing laboratories.

Local application systems are usually activated automatically by a detection system with the provision for manual operation of the system, and the majority of large systems are supplied from the dry chemical cylinder pressurized from the nitrogen or carbon dioxide cylinder.

The design of the piping for a local application system is similar to that described for the dry chemical total flooding system. Pipe lengths, pipe sizing, number and size of fittings, type, and placement of nozzles are critical. The local application system will experience the same, if not greater, velocities and pressures, requiring application of substantial pipe supports and installations to resist the extremely high thrust conditions.

Dry Chemical Hand Hose Systems

Where large fuel spill fires are a potential hazard, dry chemical hand hose lines perform as exceptional firefighting equipment. Hand hose lines are supplied from a dry chemical storage cylinder with a separate nitrogen or carbon dioxide cylinder to pressurize the dry chemical. Pressure forces it through the hose to the discharge nozzle, and the hose operator has the capability of controlling the discharge with an on-off nozzle valve. (See Figure 4.8.)

Each hose station must have a supply of dry chemicals capable of furnishing a discharge for a minimum of 30 seconds. The hose stations can be individual, self-contained units with the dry chemical supply dedicated to the individual hose station, or the hose supply can be furnished from the same dry chemical supply that is furnished for a local application or total flooding system. The supply for both the system and the hose stations is designed using selector valves that direct the discharge through the piping. The selector valve activation determines which piping system will receive the dry chemical discharge.

The flow rate of the dry chemical is a function of the hose size, hose length, and nozzle. The average flow rate is approximately eight pounds per second for a 1" hose, 15 pounds per second through a 1/2" hose, and three pounds per second through a 3/4" hose. For example, with 50' of 3/4" hose and a 3/4" nozzle, the flow rate is approximately 2.7 pounds per second, with a range of about 25', with a 150 pound supply of dry chemicals, which will discharge for approximately 49 seconds.

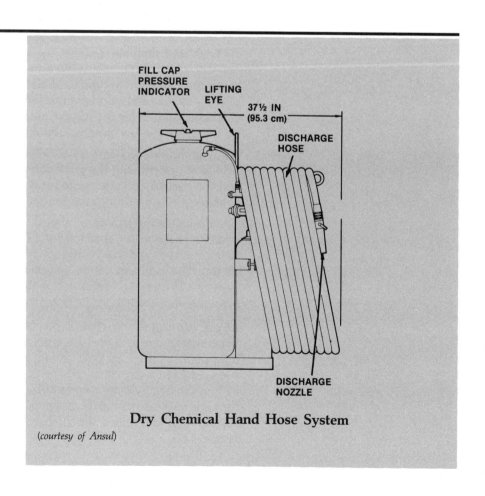

Dry Chemical Hand Hose System

(courtesy of Ansul)

Figure 4.8

The strong cylinders for supplying hose systems, and total flooding or local application systems plus hose stations, can range from 125 to 3000 pounds of dry chemical.

Pre-engineered Dry Chemical Systems

The discussion of dry chemical system has presented the many and varied difficulties in designing and calculating engineered total flooding and local application dry chemical systems. Pre-engineered dry chemical systems have already been engineered by professionals experienced in the design and calculation of dry chemical systems.

To protect relatively small hazards that have similar construction configurations, dry chemical systems are tested by the manufacturer using actual fire tests for various hazards. These tests and the test results are confirmed and verified by Underwriters Laboratories, a nationally recognized testing laboratory.

The design of the system including the number and type of nozzles, quantity of any chemical, type of dry chemical, method of operation, maximum and minimum pipe lengths, number and type of fittings, rate of application, and minimum effective time of discharge, are all part and parcel of the UL listing given by the Underwriters Laboratories. A pre-engineered system must be installed within the parameters of the UL listing which are covered in the manufacturer's installation manual. This manual is an integral part of the UL listing, and the system components listed in the manufacturer's manual can be used.

Although pre-engineered dry chemical systems may be total flooding or local application type, the most common pre-engineered systems are used for protecting kitchen hoods and hood exhaust ducts. (See Figure 4.9.) Kitchen hood systems are automatically activated by a conventional detection system or by fusible link (similar to a sprinkler head fusible link) which releases the actuating device when heat fuses the element and the link separates. There is always a manual means of discharging the system, and on some small systems manual activation is the only means of system activation. Kitchen hood pre-engineered dry chemical systems must be equipped with an automatic device to shut off fuel and power to the cooking appliances when the system is discharged.

Pre-engineered dry chemical systems are very effective in protecting laboratory hoods, paint and solvent dip tanks, and the drain boards for dip tanks, to name a few hazards. They are also an economical means of protection for hazards that can be protected with one of these UL listed systems. The sophisticated design, calculations, and actual fire testing has already been performed, and if the guidelines, mandates, equipment, devices, and design included in the UL listing are used and followed, hazard protection with dry chemicals is readily available.

Kitchen Hood System

(courtesy of Kidde)

Figure 4.9

CHAPTER FIVE
WATER SUPPLIES

CHAPTER FIVE

WATER SUPPLIES

When the municipal water pressure available is not sufficient in pressure or volume for a sustained fire extinguishing operation, then other sources must be found. If the water is available, but not at sufficient pressure, then fire pumps are used to maintain the required pressure and volume. In situations where there is no reliable water source, the use of storage tanks and reservoirs are warranted. Elevating the storage tank is one way to attain the required pressure. Another is with the use of fire pumps.

This chapter provides a basic understanding of fire pump design, devices, and equipment, and the difference between fire pumps and mechanical pumps. It also provides an overview of tanks and their use in fire protection.

Fire Pumps   Fire pumps are an integral part of fire protection, and as such, must be part of the working knowledge of anyone designing, specifying, or estimating suppression systems. This section will address the two most common fire pumps: the *diesel drive fire pump* and the *electric drive fire pump*.

Each pump manufacturer submits their pump to a nationally known testing laboratory such as Underwriters Laboratories (UL) or Factory Mutual Research (FM). These laboratories run exhaustive tests, and if the pump and all of the accessories meet their standards, the pump obtains the UL listing label and/or the FM approval. The requirements for such listings are costly, and therefore, a listed or approved fire pump is considerably more expensive than a hot water or chilled water circulating mechanical pump of equal size. Estimators, designers, and specifiers should always be certain that water supply pumps are UL listed and/or FM approved.

It is often said among mechanical engineers that the only difference between a fire pump and a mechanical pump is that the fire pump is painted red. Actually, it takes a great deal more than red paint to create a fire pump. Not only is the cost higher (the pump manufacturer invests a considerable amount in obtaining the Underwriters Laboratories and/or Factory Mutual Research tests necessary to obtain the listing or approval), but the fire pump performs with totally different criteria.

Chapter 10, on regulatory agencies and other authorities, discusses the requirements that insurance carriers place on the use of UL listed and/or FM approved fire pumps to meet the water supply volume and pressure required to maintain operable and safe fire protection systems. Insurance carrier requirements are based on how the pump is constructed, successful test results, and the performance of the pump in discharge volume at various pressures. The carrier has confidence in a pump that, except for periodic testing, may lay dormant for years, and then have to operate at 100% efficiency when activated in a fire condition.

The main distinction between laboratory testing requirements lies in the type of motor. When a fire pump is to be driven by a diesel engine, both the pump and the engine must be UL listed and/or FM approved. When a fire pump is to be driven by an electric motor, only the pump must be UL listed and/or FM approved.

General Criteria

First, we will define a few terms that identify the types of fire pumps.

Horizontal Split Case Centrifugal Pump: A fire pump that has the suction and discharge in a horizontal plane, and is identified by the pump casing, or housing, split horizontally. (See Figure 5.1.)

Vertical Shaft Turbine Pump: This is also a centrifugal type pump, but the pump is a vertical column connecting the impellers that drive the water up the column to be discharged from the top, or head, of the pump column.

Centrifugal Pump: The discharge pressure is developed by the rotation, or centrifugal force, of the water.

Basic Design Feature of Fire Pumps: A fire pump will deliver 150% of the discharge volume (gpm) it is designed to deliver, at 65% of the design pressure (psi). For example:

> A fire pump designed to deliver 2,000 gpm at 100 psi. At 3,000 gpm discharge, this pump will be discharging at 65 psi. The design gpm and psi, and the 150% gpm at 65% psi, are the two fixed points on the pump curve. There is one other fixed point on the pump curves—shutoff pressure—that reflects the gpm and psi with the pump operating against a closed valve. Shut-off pressure is sometimes referred to as the *churn pressure* because the pump is discharging against a closed valve and is just churning. Shut-off pressure cannot exceed 140% of the pump design pressure.

Horizontal Fire Pumps

"Standard" approved fire pumps have rated capacities of 500, 750, 1,000, 1,500, 2,000, 2,500, 3,000, and 5,000 gpm, and pressure ratings from 40 psi to 200 psi. "Standard" is in quotation marks because this chapter will only be dealing with these common pump capacities, although smaller capacity fire pumps are available for special uses.

Horizontal fire pumps must have a suction supply providing water to the suction of the pump under pressure. This positive pressure on the supply to the pump suction must be available when the pump is operating at 150% of its design discharge, which is called *operating at overload*.

This is an important consideration when the pump is taking its supply from a city water main. It must be carefully determined that when the pump is operating at overload, or in other words is discharging 50% more water than its design capacity, it cannot be allowed to pull the pressure of the city main below 20 psi. In addition to the volume of water being

WHEN THE CON-
TROLLER IS MOUNT-
ED ON THE BASE
THE DOOR WILL BE
ON THE RIGHT WHEN
VIEWED FROM THIS
END.

TWO ADDITIONAL 1"
DIA. HOLES FOR
MTG. BOLTS.

ANSI
COUPLING
GUARD

OUTBOARD
END

1" DRAIN
BOTH ENDS

1" DIA. HOLE FOR
4 MTG. BOLTS

VENT — .5 PIPE TAP

8.0 SUCTION
HOLES IN FLANGE
STRADDLE ₵'S

6.0 DISCHARGE
HOLES IN FLANGE
STRADDLE ₵'S

1.0 PIPE TAP
FOR DRAIN
BOTH NOZZLES

.25 PIPE TAP
FOR GAUGE
BOTH NOZZLES

CCW ROTATION VIEWED FROM COUPLING END

DISCHARGE ₵   ₵ SUCTION

CW ROTATION WHEN VIEWED FROM COUPLING END

ALL DIMENSIONS IN INCHES

Motor Frames	C	MO	HB	HF
256 T	23.1	7.1	54.0	51.0
284 TS	23.2	7.8	54.0	51.0
286 TS	24.8	7.8	54.0	51.0
324 TS	26.2	8.6	60.0	57.0
326 TS	27.8	8.6	60.0	57.0

NOTES

1. All dimensions are in inches.

2. Suction and discharge connections are 125 psi ANSI B16.1 Standard Flanges.

3. Baseplate setting (before piping), grouting procedures and final alignment must be in accordance with ITT A-C Pump recommended procedures outlined in the Instruction Manual associated with this pump type.

4. Motor dimensions are approximate for a given NEMA frame. Consult factory if space is limited.

5. Both suction and discharge pipes must be supported independently near the pump to reduce strain on the pump casing. Also expansion joints, if used, must not exert force on casing.

Fire Pump Systems

(courtesy of A-C Pump)

Figure 5.1

discharged by the pump at overload, the anticipated flow from hydrants on the city water main must also be added. The combined pump discharge volume plus the anticipated hydrant volume must not reduce the pressure in the city main below 20 psi.

The basic reason for this requirement is that the fire department pumper trucks require the 20 psi minimum pressure to operate their fire truck pumps when they take suction from the hydrant. In addition, this is a margin of safety to prevent "sucking the main dry" which could result in collapse of the main or leaking of the fittings on the main.

Suction and Discharge

The size of the suction and discharge pipe, fittings, and valves is mandated by NFPA Standard 20, and by the Factory Mutual Data Sheets. (Factory Mutual Data Sheets are equivalent to the NFPA Fire Codes.) The following suction and discharge pipe sizes are minimums:

500 gpm fire pump — 5″ suction and discharge
750 gpm fire pump — 6″ suction and discharge
1,000 gpm fire pump — 8″ suction, 6″ discharge
1,500 gpm fire pump — 8″ suction and discharge
2,000 gpm fire pump — 10″ suction and discharge
2,500 gpm fire pump — 10″ suction and discharge
3,000 gpm fire pump — 12″ suction and discharge

Note: The use of fire pumps with capacities of 500 gpm through 3,000 gpm is an example only. There are many fire pumps with various capacities up to and including 5,000 gpm, and so-called special service pumps with capacities starting at 25 gpm.

Pump Suction

The control valve in the pump suction line must be an outside screw and yoke gate (OS&Y) valve. This type of valve will provide full and unobstructed flow into the suction side of the pump with little turbulence, where a butterfly type valve might not have the vane fully open, causing an uneven flow into the eye of the impeller. It is also recommended that the OS&Y valve be installed in the upright position to further guarantee the even flow through the valve.

The one gallon per horsepower of the engine for the fuel supply is based on one pint per horsepower per hour for eight hours. The fuel supply volume should be verified by the engine manufacturer before supplying the fuel tank.

Suction and Discharge Piping

It is good engineering practice to flange all of the pump suction and discharge fittings and valves for tightness and ease in replacement or repair, although screwed and mechanical grooved joints are acceptable. The suction pipe should be galvanized to prevent tuberculation. In lieu of galvanized suction pipe, the pipe may be painted inside with an approved water-resistant paint. All discharge piping is hydrostatical tested at 200 psi for 2 hours.

Another important feature of the suction into the pump is the fact that the suction pipe size may not be the size of the flange on the pump suction. In order to reduce the suction pipe size to bolt to the pump suction flange, a flange decreasor must be used. It is imperative that the reducer be an eccentric rather than a concentric, to prevent any possibility of air being sucked into the pump and causing reduced suction and possibly creating *cavitation*. Cavitation is a complex phenomenon that can cause serious damage to the pump. The use of the eccentric reducer minimizes the introduction of air into the pump at this point of connection. (See Figure 5.2.)

The use of an elbow with the center line parallel to the horizontal pump and installed at the suction flange of the pump (or bolted to the eccentric reducer) can create turbulence in the pump and must be avoided.

When multiple fire pumps are installed, they usually take suction from a large pipe header that supplies all the pumps. If the pump suction pipe is smaller than this header pipe, it is not good practice to connect the smaller pump suction pipe to the center line of the larger pipe header. The large pipe header may not always be completely full of water, and if the surface of the water in this pipe drops below the top of the pump suction pipe, air will be drawn into the pump. If this condition could exist, then it is better to connect the pump suction line to the bottom of the large pipe header.

All of these measures—including the eccentric tapered reducer and the OS&Y valve—are designed to maintain a constant flow of water to the pump, with a minimum of turbulence, since the introduction of air and/or turbulence in the flow can reduce pump efficiency.

Circulation Relief Valve

Every pump driven by an electric motor and radiator-cooled diesel motor must have a circulation relief valve, an automatic valve that provides sufficient water circulation when the pump is operating with no flow—against a closed discharge valve. If sufficient water is not circulated in this situation, the pump would keep churning the same water in the pump casing and the water would eventually overheat and damage the pump.

In designing the pump house or pump room, a drain must be provided to carry off the water discharged from the circulation relief valve. Fire pumps driven by a diesel motor without a heat exchanger cooling system do not require a circulation relief valve because the diesel engine's water cooling

Proper Use of Eccentric Reducers at the Pump Suction

Figure 5.2

system takes suction from the discharge of the pump. When the pump is running against a closed valve, the diesel engine is obviously also running and taking water from the discharge of the pump and circulating it through the diesel engine to maintain the engine temperature. This diesel engine cooling water passes through a heat exchanger on the engine and is discharged. (This discharge must also be provided with a floor drain.)

The circulation relief valve is relatively small—3/4" or 1" depending on the size of the pump but the diesel engine cooling water could be discharging a considerable amount of water. It is best to determine from the diesel engine manufacturer how much water is anticipated before designing the drain size.

Main Relief Valve

Unless the system pressure is expected to exceed the design pressure (usually 175 psi), a relief valve is not required on fire pump systems where the pump is driven by an electric motor. System pressure could escalate if the suction pressure increased and subsequently increased the discharge pressure.

The electric motor is considered to be a constant drive motor, but the diesel engine is an *adjustable speed driver*, and also has the potential (although remote) of "running wild" and accelerating the pump's discharge pressure. A relief valve in the discharge line is, therefore, required on all diesel drive fire pump systems.

The main relief valve is sized according to the size of the fire pump—a 3" valve for a 500 gpm pump up to an 8" for a 3,000 gpm pump. To make sure there is no back pressure in the discharge pipe from the main relief valve, the relief valve discharge line is also quite large—5" for a 500 gpm pump up to a 12" for a 3,000 gpm pump.

Although the main relief valve does not discharge more than a trickle of water when the pump starts, malfunction or excess pressure could expel hundreds of gallons per minute. For this reason, the most desirable location for this main relief valve discharge is *outside* the pump house or pump room onto grade. Discharging onto the floor of the pump house or room and into a floor drain can invite damage from splashing and the accumulation of water.

If the water supply to the pumps comes from a tank where the quantity of water is limited, the main relief valve discharge can be discharged back to the tank.

Regardless of how the main relief valve discharge water is handled, the flow of water from the relief valve must be visible. This is accomplished by the use of a *relief valve cone* connected between the relief valve and the relief valve discharge line. This cone can be open or closed, with small windows to view the flow.

Pump Test

The fire pump must be tested by flowing the pump discharge after installation, and the insurance carrier may require an annual full flow discharge test. This flow test is accomplished in two ways.

1. The pump test line is connected to the pump discharge line after the discharge line check valve and ahead of the discharge line control valve.

2. The pump test line has a normally closed control valve and is piped to a test header installed on the exterior wall of the pump house or pump room.

This test header has several 2-1/2″ valves, the number of which depends on the size of the pump. (See Figure 5.3.)

Pump Test Design

When the pump is to be tested, the discharge line control valve is closed and the test line control valve opened and the pump starts up. As each of the 2-1/2″ valves open, the flow is obtained and the pump discharge pressure recorded. Subsequent valves are opened until the pump is operating at design gpm and pressure, and then on up to overload —150% of discharge and 65% of pressure.

There is another method of testing a pump. The pump test line is located as described above (in the pump discharge line), and the test line has a normally closed test line control valve. A *test meter* is installed in the test line and another normally closed test line valve installed on the discharge side of the test meter. The pump discharge control valve is closed, both test line control valves are opened, and the pump is started. By controlling the amount of water supply to the pump by gradually opening the control valve on the pump suction, the meter will indicate the flow, and the gauges on the pump will indicate the pressure up through design and overload. The discharge from the metered test line can discharge back into a storage tank, or if no other discharge area is available, it can discharge back into the pump suction.

Some insurance carriers are requiring both the test meter and the discharge header. The discharge header is used to check the calibration on the test meter.

When the test header is used, extreme care must be taken relative to the discharge, which can be the pump's capacity (150%). Hoses can be attached to the 2-1/2″ valves on the test header and run to a safe place for discharge, but the pressures and volumes encountered can cause serious damage if not carefully controlled.

This test header can be used as a hydrant if the location places it in the vicinity of a facility that requires hoses. It should not take the place of regular yard hydrants because of the time required to open valves, but it can play an important role if needed (as a supply for fire department hoses).

There are two types of relief valves: *spring operated* and *pilot operated diaphragm type*. The pilot, or pressure, relief valve can maintain a constant system pressure within very close limits as the system demands change. The spring operated relief valve is set at a predetermined pressure setting, and when this pressure is exceeded, the valve will open and relieve the pressure on the system. There should be minimal back pressure in the relief valve discharge line so as not to affect the pressure relief with the discharge.

The pipe size in Figure 5.4 for relief valve discharge lines must be increased one pipe size if there is more than one elbow in the line, if the line is quite long, or if it rises above the elevation of the relief valve in order to minimize the back pressure. In addition, there cannot be any shut-off valves in the relief valve discharge line. An unofficial "rule-of-thumb" is not to exceed 15 psi of back pressure in the discharge line of the relief valve. If this level is exceeded, pipe size should be increased. Pipe size increase may be necessary for the pump test line. If

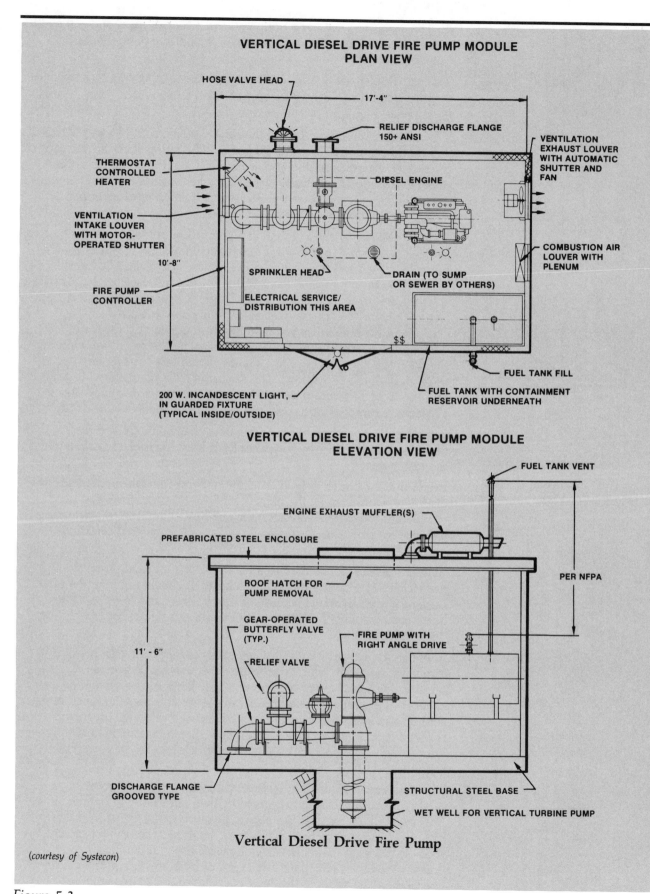

VERTICAL DIESEL DRIVE FIRE PUMP MODULE PLAN VIEW

HOSE VALVE HEAD

17'-4"

RELIEF DISCHARGE FLANGE 150+ ANSI

VENTILATION EXHAUST LOUVER WITH AUTOMATIC SHUTTER AND FAN

THERMOSTAT CONTROLLED HEATER

DIESEL ENGINE

VENTILATION INTAKE LOUVER WITH MOTOR-OPERATED SHUTTER

10'-8"

COMBUSTION AIR LOUVER WITH PLENUM

FIRE PUMP CONTROLLER

SPRINKLER HEAD

DRAIN (TO SUMP OR SEWER BY OTHERS)

ELECTRICAL SERVICE/ DISTRIBUTION THIS AREA

$$

FUEL TANK FILL

200 W. INCANDESCENT LIGHT, IN GUARDED FIXTURE (TYPICAL INSIDE/OUTSIDE)

FUEL TANK WITH CONTAINMENT RESERVOIR UNDERNEATH

VERTICAL DIESEL DRIVE FIRE PUMP MODULE ELEVATION VIEW

FUEL TANK VENT

ENGINE EXHAUST MUFFLER(S)

PREFABRICATED STEEL ENCLOSURE

PER NFPA

ROOF HATCH FOR PUMP REMOVAL

GEAR-OPERATED BUTTERFLY VALVE (TYP.)

FIRE PUMP WITH RIGHT ANGLE DRIVE

11' - 6"

RELIEF VALVE

DISCHARGE FLANGE GROOVED TYPE

STRUCTURAL STEEL BASE

WET WELL FOR VERTICAL TURBINE PUMP

Vertical Diesel Drive Fire Pump

(courtesy of Systecon)

Figure 5.3

156

the pump test line to a test header is over 15' in length, including the equivalent length of straight pipe for fittings, the next larger pipe size from that given in Figure 5.4 must be used.

If the test line for a test meter arrangement exceeds 100', including equivalent length of straight pipe added for fittings (see Figure 5.5), the next larger size pipe shall be used, and the meter shall be sized for the increased pipe size. The meter read-out device shall be sized according to the rated capacity of the pump.

When using a test header, a 1/2" ball check valve should be installed on the test header side of the control valve before the pipe passes through the wall to the test header. This ball check will automatically drain off condensation and residual water in the line and prevent freezing of the piping through the wall and in the test header.

Maximum Pipe Sizes (Nominal)							
Pump Rating gpm L/min	Suction in.*	Discharge in.*	Relief Valve in.	Relief Valve Discharge in.	Meter Device in.	Number and Size of Hose Valves in.	Hose Header Supply in.
25 (95)	1	1	$\frac{3}{4}$	1	$1\frac{1}{4}$	$1-1\frac{1}{2}$	1
50 (189)	$1\frac{1}{2}$	$1\frac{1}{4}$	$1\frac{1}{4}$	$1\frac{1}{2}$	2	$1-1\frac{1}{2}$	$1\frac{1}{4}$
100 (379)	2	2	$1\frac{1}{2}$	2	$2\frac{1}{2}$	$2-1\frac{1}{2}$	2
150 (568)	$2\frac{1}{2}$	$2\frac{1}{2}$	2	$2\frac{1}{2}$	3	$1-2\frac{1}{2}$	$2\frac{1}{2}$
200 (757)	3	3	2	$2\frac{1}{2}$	3	$1-2\frac{1}{2}$	$2\frac{1}{2}$
250 (946)	$3\frac{1}{2}$	3	2	$2\frac{1}{2}$	$3\frac{1}{2}$	$1-2\frac{1}{2}$	3
300 (1136)	4	4	$2\frac{1}{2}$	$3\frac{1}{2}$	$3\frac{1}{2}$	$1-2\frac{1}{2}$	3
400 (1514)	4	4	3	5	4	$2-2\frac{1}{2}$	4
450 (1703)	5	5	3	5	4	$2-2\frac{1}{2}$	4
500 (1892)	5	5	3	5	5	$2-2\frac{1}{2}$	4
750 (2839)	6	6	4	6	5	$3-2\frac{1}{2}$	6
1000 (3785)	8	6	4	8	6	$4-2\frac{1}{2}$	6
1250 (4731)	8	8	6	8	6	$6-2\frac{1}{2}$	8
1500 (5677)	8	8	6	8	8	$6-2\frac{1}{2}$	8
2000 (7570)	10	10	6	10	8	$6-2\frac{1}{2}$	8
2500 (9462)	10	10	6	10	8	$8-2\frac{1}{2}$	10
3000 (11,355)	12	12	8	12	8	$12-2\frac{1}{2}$	10
3500 (13,247)	12	12	8	12	10	$12-2\frac{1}{2}$	12
4000 (15,140)	14	12	8	14	10	$16-2\frac{1}{2}$	12
4500 (17,032)	16	14	8	14	10	$16-2\frac{1}{2}$	12
5000 (18,925)	16	14	8	14	10	$20-2\frac{1}{2}$	12

Summary of Fire Pump Data

Figure 5.4

Valve Supervision

The suction valve must be maintained in its open position. The open position can be stirred by installing a device that will activate an alarm when the valve closes. One method of supervision is to install a device that closes or opens an electric contact with the valve wheel. When the wheel turns, the device transmits an audible "trouble" signal to a constantly attended location.

If this electric supervision system is not practical or economical, the suction control valve can be locked open. Another method is to seal the valve open, but this requires a weekly inspection program and should only be used in the pump system if protected within a fenced area on the property that the system serves.

Pipe Size	$\frac{3}{4}$	1	$1\frac{1}{4}$	$1\frac{1}{2}$	2	$2\frac{1}{2}$	3	$3\frac{1}{2}$	4	5	6	8	10	12
Standard Elbow	2	2	3	4	5	6	7	8	10	12	14	18	22	27
Medium Turn Elbow	2	2	3	3	4	5	6	6	8	10	12	16	19	22
Long Turn Elbow	1	2	2	2	3	4	5	5	6	8	9	13	16	18
45° Elbow	1	1	1	2	2	3	3	3	4	5	7	9	11	13
Tee — 90° Flow	4	5	6	8	10	12	15	17	20	25	30	35	50	60
Gate Valve						1	1	1	2	2	3	4	5	6
Cross	4	5	6	8	10	12	15	17	20	25	30	35	50	60
Alarm Valve						10			20		18	35		
Dry Pipe Valve							10		10		19	27		
Deluge Valve					18		29		35		33			
Swing Check Valve									10		11	14		
Detector Check Valve									14		36	95	31	

Equivalent Pipe Length Chart
(Fittings and valves expressed in equivalent feet of pipe)

(courtesy of the Viking Corporation)

Figure 5.5

158

Automatic Air Release

An automatic air release valve must be installed on the casing of a split case pump to automatically release air from the pump casing when the pump starts. The automatic air release is required on fire pumps that start automatically, and shall not be less than 1/2" in size. The automatic air release must be a UL listed float-operated type device.

Diesel Fuel System

The fuel supply for a diesel drive fire pump requires at least one gallon per horsepower of the engine, a figure is based on one pint per horsepower per hour for eight hours. The fuel supply volume should be verified by the engine manufacturer before specifying the fuel tank. To this volume must be added 5% for expansion and 5% for the fuel that will remain in the sump of the fuel tank. The fuel line to the engine is located so that the fuel in the sump cannot be used by the engine.

Note: Where there are multiple diesel drive fire pumps, each engine should have its own dedicated fuel supply tank and fuel supply piping, and the fire pump diesel engine fuel supply should only serve the fire pump diesel engine. The diesel fuel tank should be located in the pump house or pump room, particularly in areas where freezing temperatures are possible. The fuel tank requires a device to determine the fuel level in the tank. A sight glass type of device is not acceptable.

No shut-off valve can be located in the fuel return line to the fuel supply tank, and the fuel supply line to the engine must be adequately protected against physical damage. The size of the fuel piping and the piping arrangement of the supply and return must comply with the engine manufacturer's specifications and requirements.

Pump Rotation

Rotation of the fire pump is important because it determines the location of the suction and discharge, and the pump rotation that is specified or stipulated on the purchase order must comply with the location of the suction and discharge indicated on the pump design. The best way to determine the direction of the pump rotation for a horizontal split case centrifugal fire pump is as follows: Imagine that you are sitting on the (driver) electric motor or diesel engine. If the suction is on your *right hand*, the pump rotation is *clockwise*. If the pump suction is on your *left hand*, the pump rotation is *counterclockwise*.

It is best to design diesel drive fire pumps with a clockwise rotation—the suction on the right. Diesel engines are available with a counterclockwise rotation on special order, but there is usually a delay plus additional cost involved. Ordering the pump with the proper rotation is critical to avoid the delay and expense of returning it.

Engine Exhaust

Each fire pump diesel engine requires a dedicated and independent exhaust system. The exhaust pipe should discharge outside the pump room or pump house, so that the exhaust discharge will not affect personnel or adjacent structures. It is important to check with the engine manufacturer regarding the type of muffler and the size of the exhaust pipe. This is especially true if there is an extensive length of exhaust pipe, to avoid exceeding the back pressure requirements of the specific engine.

Ventilation

The diesel engine manufacturer should be consulted for assurance that adequate air will be supplied to the engine. This critical item is sometimes neglected completely or grossly underestimated in the design of the pump room or pump house. The efficient operation of the diesel engine or engines depends on a supply of adequate makeup air for engine operation.

Diesel Engine Cooling

Two types of diesel engines are available that carry the UL listing and/or the FM approval for use with fire pumps. (Both the fire pump and diesel engine must be UL listed and/or FM approved). These two types are *closed circuit, liquid cooled* and *closed system type*.

The first type uses a closed circuit, liquid cooled arrangement with a heat exchanger. This liquid cooling water is taken from the discharge of the fire pump so that any time the pump is operating, water is being supplied to the engine cooling system. With this type of cooling system, the water must be discharged to waste. This means a floor drain is required in the pump room or pump house with adequate capacity to carry off the discharged water without flooding the area. The diesel engine manufacturer can furnish details on the engine cooling system, including the amount of water that will be discharging when the engine is running.

The second type of diesel engine listed for driving a fire pump is a closed system type with a radiator in lieu of a shell and tube heat exchanger. With this type of engine, a floor drain is not required, but adequate ventilation air to cool the radiator must be provided.

Reliable Power Supply to Electric Drive Fire Pumps

Reliability of the electric power supply to the electric motor driving a fire pump is extremely important, and requires careful engineering evaluation. The NFPA Standard for fire pumps and the FM Data Sheets that describe the requirements for fire pumps both provide detailed requirements for power supplied to electric drive fire pumps. The "rule-of-thumb" for power supply to electric drive fire pumps is: *two independent dedicated supplies*. When considering a fire pump installation, especially one where the fire pumps are large capacity requiring large horsepower motors, furnishing independent electric power supplies that will meet the requirements of the insurance carrier may not be cost effective. In this case it is best to evaluate the use of diesel in lieu of electric drive fire pumps. Diesel drive election is also worth considering if the fire pumps are larger than the available electric power supply.

Where two fire pumps are required by the insurance carrier—one primary fire pump and one reserve pump—it is sometimes prudent to use an electric drive fire pump for the primary pump and a diesel drive fire pump of equal size for the reserve pump. The use of a primary and a reserve pump is usually a requirement of an insurance carrier when the fire pump is the only source of volume and/or pressure to a suppression system protecting a high value property. The pumps are equally sized for volume and pressure. Should the primary pump fail to start or be out of service for repair, the reserve pump will serve as the primary pump.

Diesel Engine Exhaust

To protect personnel, the engine exhaust pipe should be encased with a high temperature insulation at least 6' above the floor, and the exhaust pipe passing through a non-combustible roof with a combustible cover should have at least an 8″ clearance between the exhaust pipe and the roof sleeve. If the exhaust pipe exceeds 15′ in length, it should be increased at least one pipe size over that recommended by the engine manufacturer.

Diesel Engine Cooling System

Water-cooled diesel engines using a heat exchanger system will discharge between 15 to 65 gpm, depending on the size of the engine. The discharge from this cooling system should be piped to an open drain located where the discharge is visible. The relief valve waste cone is an acceptable open receptacle for the discharge if the relief valve discharge line discharges to the atmosphere.

The diesel engine controller will transmit a signal indicating an overheated engine, low oil pressure, over speed, and failure of the engine to start. If overheating or low oil pressure occurs, the engine will continue to run until it fails. Some controllers have a safety feature to shut the engine down with a low oil pressure or high water temperature if this condition should occur during the weekly test. If a fire condition should occur calling for the pump start, this shutdown safety feature will be canceled and the engine will start and run to destruction.

Electric Motor Pump Driver

Diesel engines to drive fire pumps must be UL listed and/or FM approved, but electric motors to drive fire pumps do not require this listing or approval. Electric motors to drive fire pumps must be constructed to meet the National Electrical Manufacturers (NEMA) specifications, or the Canadian Electrical Manufacturers (CEMA) specifications.

Booster Pumps

When a fire pump is taking suction from a city water main where the volume (gpm) is adequate to satisfy the system demand, but the system requires more pressure than the city main can supply, the fire pump is referred to as a booster pump.

Fire Pumps in a Bypass

When the water supply to a suppression system is furnished by a water source that could provide at least minimum volume and pressure, the fire pump will be installed in a bypass. Should the fire pump be out of service, the system would have a water supply to at least control the combustion.

Pump House

The pump house or pump room should have a reliable heating system to maintain the temperature above 40°F. The temperature of a pump room or pump house for diesel drive fire pumps should not be less than that recommended by the engine manufacturer. If the temperature in the area where diesel drive fire pumps are installed is not maintained at 70°F, the use of automatic engine water heaters or oil heaters is required. These heaters maintain the engine at an efficient starting temperature and are commonly referred to as "hot starts."

Coupling Guards

The shaft between the pump driver and the pump has a coupling, and since this shaft rotates at a high rate of revolutions per minute when the pump is operating, it is necessary to have a coupling guard to ensure the safety of personnel.

Jockey Pump Controller

The pressure maintenance pump, or jockey pump, has its own dedicated controller which automatically starts the pump when the system pressure drops below a predetermined point, and automatically shuts the pump down when the required system pressure has been restarted.

Controller

The controller for diesel or electric drive fire pumps is completely assembled, equipped, wired, and tested at the factory before it arrives on site. For this reason, it is imperative that no alterations or additions be made to the controller after it leaves the factory, as this will void the UL listing and/or FM approval. Each fire pump shall have its individual control panel, and the control panel should be installed within sight of the pump it controls. The starter switch for an electric motor drive pump or diesel drive engine pump is contained within the control panel. Most fire pumps start automatically, and are stopped manually, with the starter switch actuated by a drop in water pressure. When a suppression system is activated, sprinkler heads fuse, a signal from a detection system trips a valve, and there is a drop in the system pressure as water begins to flow. A small, 1/2" minimum size water line is connected to the system piping on the discharge side of the check valve in the pump discharge line and piped to the pressure switch in the controller. The water pressure in this line reflects the system pressure drop and subsequently trips the pressure switch which, in turn, starts the pump motor or pump engine. The constant static pressure is maintained in the system piping by a small *pressure maintenance pump*, commonly referred to as a *jockey pump*. The jockey pump probably derives its name from the fact that it "rides" on the system. The jockey pump takes suction on the water supply side of the fire pump suction control valve so that it has a water supply even if the suction control valve is closed. The suction side of the jockey pump has an indicating type control valve, and the discharge line from the jockey pump has a check valve and an indicating type control valve. These control valves are intended to facilitate repair of the jockey pump. The discharge of the jockey pump is connected into the system piping on the system side of the fire pump discharge check valve and control valve. Jockey pumps should have very small discharge volume, especially on tight systems (such as sprinkler systems) because there should not be any leakage. If the jockey pump is discharging into large multiple systems supplied by underground piping, the discharge should be sized at least to compensate for possible leakage.

Pressure Considerations

The pressure rating of the jockey pump should be 10 to 15 pounds above the starting pressure of the fire pumps. The fire pumps should never be used as pressure maintenance pumps, and the jockey pump should not have a discharge volume that will satisfy any of the suppression system demands. Fire pumps can also be started on an electric signal from a suppression system, with the system water pressure drop starting as a backup. Automatic starting of a fire pump by activation of a water flow device in lieu of water pressure can be used in situations where the water pressure of the supply fluctuates so much that a predetermined cut-in pressure starting cannot be maintained. Another instance where the pumps may be started by a signal from a detection system or the activation of a water flow device is when the suppression system is protecting a hazardous area and design pressure on the system is needed quickly. The protection of an aircraft from a fuel spill fire is an example.

A few seconds delay can mean the destruction of an aircraft involved in a jet fuel spill fire. The fire pumps, especially the diesel engines that can take a few seconds to come up to speed, should be started to bring demand pressure on the system the second a foam deluge or foam monitor nozzle system commences discharge. When a sprinkler system protects a hazardous area, a fire may almost immediately open numerous heads, and the need for immediate water pressure is imperative. When it is important to start the fire pumps before waiting for the system pressure drop, it is necessary to have the pressure drop starting as a backup.

Controller Cabinet

As mentioned previously, the controller cabinet contains the motor or engine starter, the circuit breaker and the disconnect switch, the manual start device, and the manual stop device. It also provides the relays for alarm and control devices, plus annunciation of the condition of the pump and driver. The controller should be mounted on a dedicated concrete base not less than 12" above the floor to prevent any water from the pumps causing damage to the controller. In designing the location of the controller in the pump room or pump house, the dimensions for the controller being used can be obtained from the manufacturer. If the cabinet is serviced from the rear of the cabinet, allow at least 3-1/2' from the back of the cabinet to the wall, and to permit inspection and servicing, allow not less than 2' on the sides of the cabinet.

Controller Alarm and Signal Devices

Electric drive pump controllers have a power available pilot light on the panel. If the pump room or pump house is not constantly attended, remote audible and visible signals shall be transmitted to a constantly attended location. These signals are:

- Controller has started the motor and the motor is in a motor running condition.
- Loss of power to the motor starter.
- Phase reversal on the line side of the motor starter.

Diesel engine drive fire pump controllers have a visible indication that the controller is in an automatic condition. The diesel engine controller should also have the following visible and audible alarms. If the pump room or pump house is not constantly attended, these signals should be transmitted to a constantly attended location:

- Low oil pressure
- High engine jacket coolant temperature
- Failure of engine to start automatically
- Engine shutdown from overspeed
- Battery failure
- Battery charger failure

Some controllers are required to have a relay that will start the pump motor or pump engine when there is a power failure to the controller. With electric motor drive pumps, if the power supply is adequate, the motor should start "across-the-line" to full power on signal from the starter. If the power supply is not capable of accepting the surge of power, the motor can use a reduced voltage starting sequence where the motor comes up to full speed slowly or in stages. When using multiple electric drive pumps, the motors should be started in sequence, with a several second interval between the starting of successive motors to prevent a breakdown of the power source because of the heavy load.

Weekly Diesel Drive Pump Test

The controller for a diesel drive fire pump must be equipped with an automatic start of the engine once a week, and a device to keep the engine running for not less than 30 minutes. The controller can be equipped with an automatic shut-down. This weekly test run is recorded on a pressure recorder.

Diesel Engine Batteries

The diesel pump starts from not less than two batteries that are mounted on the engine base. The batteries are constantly charged by a battery charger that is specifically listed for use with fire pumps.

Vertical Turbine Fire Pumps

When it is not possible to have a water supply to the pump under a positive pressure, the use of a vertical turbine fire pump should be considered. A vertical shaft pump is a centrifugal pump with the impellers in a column. The primary consideration with vertical fire pumps is that the source of water is adequate and dependable. This source can be from reservoirs, lakes, or rivers supplying wet pits. Vertical turbine fire pumps can be driven by an electric motor mounted on the top of the pump shaft, or by a diesel engine operating the pump shaft through a right angle gear. The design of the supply to a vertical turbine fire pump is rather sophisticated, and should be calculated and designed by experienced fire protection engineers. The support of the pump shaft and electric motor or right angle gear drive is a consideration of a professional structural engineer. The pump house design must incorporate a closable opening in the roof to allow the installation and removal of the pump shaft. The electric motor for a vertical pump must be equipped with non-reverse ratchets, and the right angle gear used when the pump is driven by a diesel engine must be equipped with a non-reverse ratchet.

Miscellaneous

When pump manufacturers submit quotations for a fire pump project, the difference in costs may be a reflection of what is included in the quote. It is important to stipulate in the bid request exactly what equipment should be included in the quotation.

The following is a list of items the pump manufacturer can and should furnish:

- Pump
- Driver
- Controller
- Jockey pump and jockey pump controller
- Flow meter easer
- Hose test header
- Relief valve
- Closed waste cone
- Open waste cone
- Circulation relief valve
- Eccentric reducer
- Concentric increaser
- Pressure recorder
- Relief valve tee

Flushing and Testing

Before performing the fire pump acceptance tests, the piping must be hydrostatically tested at 200 psi for two hours with no leaks. If the system pressure is to be greater that 150 psi, it will be necessary to determine the maximum system pressure and establish the hydrostatic test pressure at 50 psi over the maximum system pressure. The fire pump suction piping must be flushed at a predetermined flow rate. Once the system demand has been determined by hydraulic calculations, the flushing flow rate shall not be less than the calculated system flow rate. It is possible that the calculated system flow rate will not be sufficient to meet the flushing flow rate requirements of the NFPA, which are based on the pipe size of the suction piping.

The pump has a 10″ suction pipe and the calculated system demand is 1800 gpm. A check of the NFPA Standard for fire pumps, Standard 20, indicates that the flushing flow rate for a pump with a 10″ suction is 2440 gpm. NFPA 20 will indicate the greater flow rate system demand gpm or NFPA flow rate requirements. When the flushing and testing has been completed, the contractor will provide a certificate verifying the successful completion of the flushing and testing.

Fire Pump Field Acceptance Test

When the fire pump installation is subjected to the field acceptance tests, the following representatives must be present:

- Pump manufacturer
- Engine manufacturer (if diesel drive)
- Controller manufacturer
- Transfer switch manufacturer (if supplied)

These representatives include the cost of attendance in the quotation or at least have some agreed financial arrangement with the contractor. The owner, insurance carrier, engineer, and all authorities having jurisdiction (or their appointed representatives) should be present during the tests. Parties should be notified of the place, date, and time of the tests not less than five days before the test date. There is always a concern that because the pump water supply is not adequate to allow testing the fire pump at overload, or 150% of its design, the test will not be successful. If this situation exists, the pump shall only be required to discharge the maximum that the supply is capable of furnishing to complete this phase of the test successfully. The installing contractor will furnish all labor, material, and devices required for the tests. It is recommended that the contractor run the pump through the test procedure a few days before the official field acceptance test date, to uncover any deficiencies and make the necessary corrections or repairs. The fire pump will be required to discharge at the minimum, the rated, and the peak pump loads, and the results will be compared with the manufacturer's certified pump test curves. The controller manufacturer has a recommended test procedure, but it is important that this procedure include a minimum of 10 automatic and 10 manual pump starts, and that the pump driver operate at full speed for not less than five minutes during each of these test starts. All alarms, both local and remote, will also be activated and tested. Following completion of the acceptance tests, the fire pump installation should be returned to the automatic operating condition, including refilling of the diesel fuel tanks. The NFPA Standard 20 and the FM Data Sheets contain information regarding field acceptance tests. In addition, the insurance carrier may have specific acceptance test requirements. It is recommended that the insurance carrier input be obtained relative to their acceptance test requirements. The preparation of a test schedule will expedite the test procedure, especially if the acceptance test involves a multiple pump installation.

Fire Pump Controller Alarm Signal Options

The following options can be provided as relays in the controller to transmit trouble signals to a constantly attended location. Some of the following may be required by the insurance carrier:

- Low pump room (house) temperature
- Relief valve discharging
- Flow meter left on by passing pump
- Suction supply water level below normal
- Suction supply water level near depletion
- Diesel fuel supply below normal

Tanks If the pressure demand of the suppression systems exceeds the city main's capacity, the first consideration is the use of a fire pump to boost the pressure. If, on further examination (See Chapter 8, "Hydraulic Calculations"), it is found that the booster pump can furnish the required demand pressure, but at the volume (gpm) demand of the suppression systems, the city main pressure will be below the 20 psi residual pressure necessary to allow the fire department pumper trucks to utilize the hydrants. On the other hand, the supply main may not be large enough to furnish the volume demand of the suppression systems. In either case, the use of a stored water source to supply the fire pump and system demands is warranted. For example:

> A large company builds a large manufacturing and rack storage warehouse in a small town. Because of the extensive rack storage system, the demand for the suppression systems far exceeds the supply offered by the city water system. The result is a storage tank dedicated to supplying the fire protection filled from the city water system, and furnished to the suppression systems through the fire pumps. Actually, two storage tanks of equal volume are required, each dedicated to fire protection supply. This requirement is based on the concern that if one tank were out of service for repair and painting, the facility would be without fire protection. This is a very important consideration in recommending and designing reliable fire protection water supplies.

Tank Selection

Water storage for fire protection is available in ground level steel tanks, elevated storage tanks, and ground level reservoirs. The ground level steel tanks are referred to as *pump suction tanks*, and are available in a wide range of capacities. Approval of the fire tank system must be obtained from the insurance carrier. Factory Mutual has a list of approved tank designs, capacities, and dimensions, and the data sheets mandate the construction and installation of these tanks. NFPA's Standard number 22—Water Tanks For Private Fire Protection—mandates the construction, design, and installation of water storage facilities for fire protection system water supplies. If the client is insured by Factory Mutual, the system must use FM approved tanks and installed in accordance with FM standards. If the client is insured by an insurance carrier other than FM, the specifications, cost estimates, and purchase orders must stipulate that the tanks conform to the requirements of NFPA relative to construction, design, and installation.

To lend some flexibility to these statements, a manufacturer of Factory Mutual approved water storage tanks has the option to design, fabricate, and erect a tank using dimensions other than those that have been approved by FM, but the tank is subject to review and approval by

Factory Mutual Research. FM Research is part of the FM insurance family that establishes the specifications for FM approved devices, materials, systems, and equipment, and lists them in the FM Approval Guide. In the Factory Mutual Approval Guide, the dimensions—both height and diameter—for each tank capacity are presented. Steel-welded FM approved pump suction tanks vary in capacity from 100,000 to 1,000,000 gallons, and steel bolted tanks vary in capacity from 30,000 to over 600,000 gallons. Factory Mutual also lists approved elevated gravity tanks in standard sizes from 100,000 to 500,000 gallons, on 100 to 135 foot steel towers. NFPA Standard 22 also details dimensions and capacities for a variety of tanks, including wooden elevated gravity tanks or wooden towers. Although some yard or site conditions dictate the dimensions of a ground level pump suction tank, or the height, dimensions, and location (possibly on a building in lieu of a tower) of an elevated gravity tank, if at all possible the use of an FM approved standard tank design and size will save the trouble, time, and expense of obtaining approval of a special design. In geographic areas subject to freezing, these tanks require an approved form of freeze prevention.

Concrete Reservoir

If space is available, a concrete ground level reservoir can be provided for fire protection water. These reservoirs come in all sizes and shapes, covered or open. The suction originates from a point in the reservoir that provides a positive suction pressure to horizontal pumps.

The pit where vertical turbine pumps are employed must be designed very carefully to allow adequate room around the pump shaft to prevent the water flow from forming a vortex, which could affect the efficiency of the pump. Webster defines a vortex as a "whirling mass of water which forms a vacuum at its center." It is easy to visualize the affect on the efficiency of the pump with the pump shaft in the center of a vortex. A mini-vortex is the whirling water rotating around the open throat of a drain. In a wet pit, the pump shaft with the impellers is rotating at tremendous revolutions per minute.

The pump manufacturer should provide the correct dimensions of the wet pit for their specific vertical turbine pump.

In pump suction tanks, the suction pipe is installed near the bottom of the tank, with an elbow on the end of the suction pipe in the tank. To avoid the creation of a vortex as the water rushes into the suction pipe elbow, a vortex plate is fastened to the elbow.

The vortex plate fastened to the suction elbow is a 4' by 4' square plate, located a distance one half the diameter of the suction pipe above the tank floor, but not less than 6" above the tank floor. Water entering the suction elbow must pass around the square sides of the vortex plate, which prevents the swirling or whirling action of the water entering the elbow. A single bare end suction pipe would create the same action as the water entering the open drain, accented because this water is being sucked into the suction pipe in great capacity and with considerable velocity, and a vortex would surely form. Where yard site space was limited, concrete reservoirs have been constructed under the floor of the pump room or beneath the basement floor of a building, using vertical turbine pumps to supply water to the suppression systems. It was previously noted that two above ground steel pump suction tanks were required to be installed for reliability of the water supply. With concrete reservoirs, it is only necessary to split the reservoir into two sections for the same results. If one side of the reservoir should be drained for repair or cleaning, the other side of the reservoir would still have adequate water supply to meet the system demand. The piping supplying the pumps from this divided reservoir must take suction individually from both sides of the reservoir with control valves in each suction pipe, and the two pipes connected to a common pump suction header. This design allows the pumps to take suction from either side of the reservoir, or both sides simultaneously.

Embankment Fabric Reservoir

A reservoir that deserves serious consideration as a fire protection water supply is an embankment-supported, rubber-coated, fabric tank. (See Figure 5.6). The tank is actually a reservoir liner with a cover. When the liner is filled with water, its top floats upward and remains floating when the tank is full of water. The earth banks are built on all four sides of a square, and the tank liner is laid, unrolled, in the square within the embankments. Once the liner is in place within the open enclosure, it is anchored to the embankments and the piping connections are completed. The actual installation of the tank, or reservoir, liner requires about one day. The tank is constructed of a high-strength nylon fabric that is coated with synthetic rubber that not only seals the liner water-tight, but also protects it from deterioration from weather, abrasion, temperature changes, chemical fumes, and ozone. The synthetic rubber coating keeps the liner flexible and shock-resistant, and therefore the tank is able to resist severe damage in earthquake areas. The inlet-outlet drain fitting is installed in a concrete pad with a vortex plate bolted to the inlet-outlet drain fitting. The exposed surface of the filled tank is painted as recommended by the manufacturer, and is applied according to the recommendations. In areas subject to freezing, the tank water should be heated by circulating the water through a heat exchanger. These embankment tanks are available in capacities of 25,000 to 1,000,000 gallons, and range in size from 33 square feet (25,000 gallons) to 128 square feet (1,000,000 gallons).

Tank Heating

Embankment reservoirs, like all the tanks and reservoirs described, must be constructed, installed, and heated (in areas subject to freezing), in accordance with the requirements of NFPA Standard 22 and the Factory Mutual Data Sheets. The map in these publications indicates (by means of temperature lines) the lowest one day mean temperature for all areas of the country. By locating the tank site on this map, the lowest temperature that the tank will experience establishes the amount of heat that will be required. This determination of the lowest one day temperature the tank will experience, and the capacity of the tank, are both factors that are used in calculating the amount of heat necessary to maintain the water temperature in the tank at or above 42°F. A low water temperature alarm should be provided to alert personnel should the heating system fail to maintain this minimum temperature of the water. The heating arrangement and equipment should automatically control the water temperature in the tank.

Tank Water Level

Tanks and reservoirs should be kept full automatically by use of an altitude valve arrangement or some other approved configuration. All tanks should have some device to determine their water level, and a low water level alarm device should the water level fall below normal capacity.

Nominal Capacity (gals.)	Tank Dimens. (ft.)			Dike Dimens.		Average Earth Dike (cu. yds.)	Tank Net Weight (lbs.) (Approximate)
	A	B	C	D	E		
25,000	33.2	15.2	6	22.5	1.5	549	836
50,000	41.5	17.5	8	29.0	2.0	1045	1,286
100,000	52.4	22.4	10	35.5	2.5	1855	2,045
200,000	66.4	30.4	12	42.0	3.0	3137	3,269
300,000	77.7	41.7	12	42.0	3.0	3628	4,414
400,000	87.2	51.2	12	42.0	3.0	4041	5,506
500,000	95.6	59.6	12	42.0	3.0	4402	6,563
600,000	103.1	67.1	12	42.0	3.0	4728	7,597
700,000	110.0	74.0	12	42.0	3.0	5028	8,615
800,000	116.5	80.5	12	42.0	3.0	5307	9,617
900,000	122.5	86.5	12	42.0	3.0	5568	10,609
1,000,000	128.2	92.2	12	42.0	3.0	5816	11,593

Note: A= Tank Top–Square B= Tank Bottom–Square C= Tank Depth D= Earth Dike–Base E= Earth Dike–Top

EMBANKMENT AND TANK CROSS SECTION

Embankment Sizes and Data

Figure 5.6

Summary Tank and reservoir designs should be engineered by experienced professionals and involve all concerned disciplines—civil, structural, mechanical, and electrical. The location of the tank requires the expertise of a civil engineer, the design of foundations and structure of a concrete reservoir mandates nothing less than the engineering expertise of a structural engineer. The piping and heating equipment, design and engineering, should only be trusted to a mechanical engineer, and all alarms and control devices and equipment can only be engineered and designed by an electrical engineer.

The fire protection engineer can determine the required capacity of the tank, and select the appropriate, cost effective, water storage unit. He or she can design and engineer, and specify the required fire protection features, all based on experience and expertise and to meet the requirements of NFPA and/or FM and the insurance interest having jurisdiction.

CHAPTER SIX
EQUIPMENT AND DEVICES

CHAPTER SIX

EQUIPMENT AND DEVICES

The proper location of valves and other critical devices is important to ensure that the system may be shut off promptly after sprinklers have extinguished the fire, or to conserve water and pressure if pipes are broken. Equipment such as strainers are necessary to prevent clogging of the nozzles. These devices, as well as backflow preventers, tap connections, and hangers, are integral parts of any fire prevention system. The purpose, use, and installation of these important elements are discussed in this chapter.

Control Valves

Every suppression system must have a control valve, with one requirement—the valve must visibly indicate whether they are open or closed.

The Outside Screw and Yoke

OS&Y Valve is one of the most common and easily recognizable fire protection system control valves (Figure 6.1). This valve has a threaded stem that projects several inches above the operating wheel when in the open condition. When closed, the valve's threaded stem retreats into its yoke, and the top of the stem becomes flush with the wheel; thus its name, OS&Y is derived from its operation. The *outside screw* rises above the yoke when it is open, and the *yoke* is part of the body of the valve where the stem lodges when the valve is closed.

The Post Indicator Valve (PIV)

This is one of the most common fire protection valves, controlling a system from outside the building. The *post indicator* portion of the valve is a post that protrudes approximately 3' from the ground with a "window"—an open slot in which the word "open" or "shut" appears. (See Figure 6.2.) A cast steel wrench is attached to the top of the post that is designed to house a padlock for security. A wrench fits the nut on the top of the valve stem, and when the stem is turned to close the valve, the window will read "shut".

When the PIV is installed on the underground piping, the valve part is a *non-rising stem type valve* (N.R.S.). This means that when the valve is opened, the stem does not rise vertically as it does in the OS&Y valve. When the valve is opened or closed, the stem turns within the post indicator portion of the PIV.

When PIV valves control the water supply to a suppression system, they serve two main functions. First, the control valve is *closed* when making repairs on the system or when remodeling or adding to the system. The second primary function of the control valve is to *shut the system down* in a fire situation. For example, a fire has gotten out of control and opened so many sprinkler heads that the water supply pressure at the hydrants has dropped below the fire department's requirements. Another example might be that a roof may collapse due to the effect of heat from a fire on the structural steel; the collapse may break a large sprinkler supply pipe; or an explosion destroys the sprinkler supply piping.

Any of these incidents would create a tremendous loss of fire protection water not directed toward extinguishing the fire. This loss causes decreased pressure and volume at the fire hydrants, that seriously hinders firefighting. If the control valve on the system is an OS&Y valve located inside the building, in all probability it would not be accessible. A similar situation may occur if the fire approaches the system riser and control valve location, because the fire could prevent access to the control valve. This is the primary reason that the authority having jurisdiction over approval of fire protection plans and installation will require a PIV, located *outside* the building for control of the system water supply. There are some extenuating circumstances that will allow the control valve to be an *indoor* OS&Y valve:

If the use of a PIV is not possible due to site conditions;
1. The system risers with the control valve are located close to a main door in the outside wall, and
2. The outside wall is without windows for the fire to break through and prevent access through the main door.

If these two conditions exist, it may be possible to use an indoor control valve. If the system riser and control valve are to be located in a basement, however, the fire department and/or the insurance carrier will require an outdoor control valve.

STEM TYPES
(OS&Y)—Rising Stem—Outside Screw and Yoke
Offers a visual indication of whether the valve is open or closed. Recommended where high temperatures, corrosives, and solids in the line might cause damage to inside-valve stem threads. The stem threads are engaged by the yoke bushing so the stem rises through the hand wheel as it is turned.

Outside Screw and Yoke Valve

Figure 6.1

The ideal location of the PIV is about 40' from the building, to allow access to the control valve even if a fire breaks out of the exterior wall or an exterior wall collapses. Despite this requirement, there are many situations where it is impossible to locate the PIV 40' from the building, or where use of a PIV is not possible (when there is no yard around the building, or the entire area is congested with roads). There is an alternative—the use of a *wall type post indicator valve*. The wall type indicator post valve's post is installed on the exterior of the building and the valve shaft passes through the wall, connected to the non-rising stem valve installed in the system riser.

There are three types of wall posts, with all three operated with a *wheel* in lieu of a wrench, and all three have the window shut/open indicator. One type of wall post extends about 14" from the face of the wall, one type has a post extending up to 52" from the wall; and the third is completely recessed in the wall in a cast iron frame.

Important Considerations When Using the PIV

When ordering the post indicator valve for underground piping, the distance from the ground level to the bottom of the main is important. Naturally it is imperative to state the size of the main to obtain the proper size valve, but the depth of the underground pipe trench, which is the bottom of the pipe, is also critical because the post extension that connects the post to the valve is available in various lengths. This dimension must be fairly accurate or the post will either protrude too far above ground (unsightly) or too close to the ground (unseen).

The rotation of the valve opening is another consideration—shall the valve open by turning right or left, clockwise or counterclockwise? The new valve should open the same way that other valves on the property open, in order to avoid confusion in a fire condition. Should a valve be forcefully turned the wrong way, it could damage or possibly snap the valve stem. Unless otherwise specified, the PIV will turn in a counterclockwise opening direction.

For wall type post indicator valves, it is important to designate the center line of the riser where the non-rising stem valve is located. The wall type indicator is manufactured in three categories. Type 1 and Type 3 require a concrete or concrete block wall with enough thickness to accept the frame for the fully recessed post indicator, or enough thickness to provide a sleeve for the valve stem. If the wall is a thin wall, the Type 2 wall post indicator is appropriate.

Post Indicator Valve Assembly (PIVA)

One post indicator valve on the market is an assembly—a single unit containing post indicator and valve. The post indicator with the target window is separate from the non-rising stem valve installed in the underground piping. The post indicator is bolted to the non-rising stem valve. The PIVA has a butterfly valve permanently attached to the post indicator. Instead of an indicating window, the PIVA has a round hole that is open when the valve is open and completely closed when the valve is closed. In lieu of the wrench to open and shut the valve, the PIVA has a detachable crank. One advantage of the PIVA is that the valve will remain open should the post be severely damaged, even to the point of being sheared off. Another advantage is that without regular operation or exercise, the old style post indicator can be extremely difficult to operate after several years. The PIVA is designed to be operated rather easily even years after installation.

Butterfly Valve

The indicating butterfly control valve is a popular substitute for the large OS&Y valve. This is a thin valve which does not have the long yoke and stem of the OS&Y valve. Therefore, it does not take up valuable installation space both in and around the perimeter of the system riser. For example:

Installed in a system riser as the control valve, the flanged butterfly valve takes out 5″ for a 6″ riser, and 6″ for an 8″ riser. By comparison, a flanged OS&Y valve takes out 10-1/2″ when installed in a 6″ riser, and 11-1/2″ in an 8″ riser. The center line to the outside of the handle of a butterfly valve measures approximately 11-1/2″ for a 6″ valve, and 12-1/2″ for an 8″ valve. The 6″ OS&Y valve, center line to the tip of the stem in the open position, requires almost 32″ for a 6″ valve, and almost 41″ for an 8″ valve.

Post Indicator Valve

Figure 6.2

The example compares flanged butterfly vales with flanged OS&Y valves, but when the butterfly valve is designed to be used with grooved couplings or fittings, the thickness is much less than that of the flanged butterfly valve. In order to qualify as an indicating type valve, the butterfly valve has an arrow indicating the open or shut condition. The butterfly valve is available with either a wheel or crank operator and can be provided with supervisory switches to signal a trouble alarm should the valve be closed. The physical dimensions and weight difference also make the butterfly valve much easier to install, cutting down on labor cost. (See Figure 6.3.)

Valve Pit OS&Y and Post Indicator

If a building or facility has multiple fire suppression systems, and they are all supplied by a common below ground fire main, it is good practice to provide a single valve pit in the main line that feeds the fire suppression systems. This arrangement is usually a requirement of the code authority to prevent the fire department pumper from being connected to the wrong building siamese connection, rather than to the building needing protection.

The valve pit has a yard type fire department pumper connection that sits on top of the concrete pit. When the fire department pumps into this siamese, they are pumping into the entire underground piping system and into whichever building system is operating. (See Figure 6.4.)

Butterfly Valves
Butterfly valves provide bubble-tight closure with excellent throttling characteristics. They can be used for full-open, closed and for throttling applications.
The Butterfly valve consists of a disc within the valve body which is controlled by a shaft.
In its closed position, the valve disc seals against a resilient seat. The disc position throughout the full 90° rotation is visually indicated by the position of the operator.
A Butterfly valve is only a fraction of the weight of a Gate valve and requires no gaskets between flanges in most cases.
Recommended for frequent operation and adaptable to automation where space is limited.
Wafer and Lug type bodies when installed between two pipe flanges, can be easily removed from the line. The pressure of the bolted flanges holds the valve in place.
Locating lugs makes installation easier.

Butterfly Valves

Figure 6.3

The concrete valve pit contains:

- an OS&Y valve
- a check valve to prevent the fire department from pumping back through the system water service
- a second OS&Y valve
- the tee for the fire department pumper connection (which also has a check valve in the line to the siamese).

If a post indicator valve is required underground, it is possible to use a non-rising stem valve in lieu of the second OS&Y valve and, with an extension up through the top of the concrete pit, install a post indicator mounted there. Valves are necessary on either side of the check valve so that they could be closed, and the check valve removed for repair or cleaning. With the connection for the fire department pumper on the house side of the control valve, when both valves are closed and the check valve removed, the fire department may still use the pumper connection to supply water under pressure to the suppression systems.

Check Valves

Underwriters' Laboratories and/or Factory Mutual-approved swing check valves provide a minimum of friction loss. The check valve, disc, or clapper swings entirely clear of the flow of water. (See Figure 6.5.) There is now available a compact *wafer check valve*. This is smaller than the *iron body check valve*. For example:

An iron body check valve measures 17″ face to face of the flanges, and an 8″ iron body check valve measures almost 21″ for the same. By comparison, a 6″ wafer check with grooved ends measures a little over 9″ end to end. The disc on the larger iron body check valves is usually cast iron, and the wafer check valves usually have spring-loaded bronze discs.

Wafer check valves are not only compact, but weigh much less than the conventional swing check and, therefore, are labor saving as well as materially cost effective.

Detector Check Valves and Full Flow Fire Meters

Some cities require that the fire water supply to suppression systems be metered. Originally this was to keep citizens from stealing water from the unmetered fire supply. Although extremely difficult without operating all the alarms or energizing the fire pumps, this "theft" has occurred. For cities that do require metering, the cost of water used to fight a fire is not their concern because as much, if not more, water would be used by the fire department using the unmetered fire hydrants. Nonfire usage should be metered.

It was out of this concern that the *detector check valve* was developed. This valve is basically a regular iron body swinging disc check valve, with the disc, or clapper, weighted. Tapped into each side of the check valve around the disc is a small pipeline with a water meter. When there are small flows in the fire main, the weighted clapper will not rise off its seat and the water flow, seeking the line of least resistance, passes through the small pipe bypass and into the meter. When a suppression system is activated and there is a fire flow in the main, the demand flow of the suppression system raises the clapper, allowing a full flow through the check valve. With this detector check valve, only small flows of water are metered; fire flows are not.

The concept is that with a suppression system, the fire would be controlled and extinguished in a relatively short time, with the fire department using unmetered public water from the fire hydrants just

Typical Yard Fire System for a Large Industrial Plant

Figure 6.4

Item	Description	Material
1	Body	Ductile Iron
2	Cover	Steel
3	Screw	Zinc Plated Screw
4	Cover Gasket	S.B.R. Fiber
5	Hinge Pin	Stainless Steel
6	Retaining Ring	Stainless Steel
7	Clapper	Stainless Steel
8	Clapper Rubber	E.P.D.M.
9	Retainer Plate	Stainless Steel
10	Screw	Stainless Steel
11	Nut	Stainless Steel
12	Seat	Bronze
13	Pipe Plug	Cast Iron

Swing Check Valve Model D

(courtesy of Viking Corporation)

Figure 6.5

for mop up. The detector check valve may also be used in the supply line to serve as a regular check valve, and must be UL listed and/or FM approved.

Key Valve

A key valve is a non-rising stem valve installed in the underground piping. Covering the operating nut on the top of the valve is a round metal shaft known as a *roadway box* that fits over the top of the valve and extends to the surface of the ground or pavement. A metal cap fits on the top of the roadway box, and cast in the metal of the cap is the word "water". To operate this valve, the cap is removed and a long handled metal wrench is slipped into the roadway box and fitted over the valve operating shaft. The operating wrench, or "key", has a crossbar to facilitate turning and opening or closing the valve. Since the key valve is not an indicating valve, it is never to be used as a control valve on water supplies to suppression systems.

Fire hydrants require periodic maintenance. Those on private property are supplied from the fire main that feeds the suppression systems. Key valves are installed on the supply to the fire hydrant so that when the hydrant requires maintenance, the key valve can be closed, eliminating the necessity of shutting down the entire fire main. Hydrants are important firefighting equipment, and it is much more practical to have one hydrant out of service than the whole system shut down for repair.

All valves must have the UL listing and/or FM approval. This requirement applies to post indicator and non-rising stem valves, as well as the entire PIVA assembly.

Full Flow Fire Meters

When a meter is necessary for both small and full flows in the fire line, a full flow fire meter must be installed. These meters are large and expensive compared with the detector check meter, and are installed between two OS&Y valves. Full flow fire meters may measure approximately 45″ for a 6″ line, and 53″ for an 8″ line. The code authority having jurisdiction may also require a bypass around the full flow meter. Should the meter be out of service, the normally closed bypass would be opened.

When a bypass line is required, the overall measurement of the meter and bypass connections increases 98″ for a 6″ line, and 112″ for an 8″ line—face to face of the flanged bypass tees. Full flow meters are required by city ordinances, when the fire line is used to fill a reservoir or when testing deluge systems. The requirements of the local code authorities will determine their needs for metering the fire water lines. Most municipalities and authorities do not require metering, but if this is not determined up front and is later discovered, it can be a costly addition to the system, especially if full flow metering is required.

Backflow Preventers

Fire protection water in sprinkler systems may lay dormant for years and, as a result, can become contaminated. A pendent head that has been installed for a number of years, untouched, contains water trapped in the drop. This contaminated water might find its way back into the potable drinking water if the sprinkler system is connected to a public water main and a flow reversal occurs. Backflow preventers, installed between the potable water supply and the sprinkler system to prevent such contamination, are not only becoming more popular, but have become mandatory in most states and cities.

There are two basic types of backflow preventers: the *double check valve arrangement* (Figure 6.6) and the *reduced pressure type* (Figure 6.7). The installer is not given a choice of which type to use; state or local requirements specify the type. Both types must always be installed between two shut off valves, preferably OS&Y valves.

The double check valve backflow preventer does not use standard check valves. They are spring loaded check valves, and the entire arrangement is UL listed. The double check valves in this type of backflow preventer can be *detector check type* with a small metered backflow line around the preventer. Some models of backflow preventers can be installed in the vertical position, and others horizontal only. The backflow preventer should be installed immediately ahead of any system risers where the underground supply line enters the building. The reduced pressure backflow preventers require provisions to facilitate the use of test connections. Backflow preventers must be UL listed and/or FM approved.

Tapping Connections

There are two ways to make a connection to an existing, active, underground fire main: the *dry tap* or the *wet tap*. The dry tap is accomplished by shutting off the water supply to the primary, or isolated section of a main, where the connection is to be made. The pipe is drained, cut, and a section removed. A sleeve and the new connections

Reduced Pressure Backflow Preventer

Figure 6.6

Double Check Arrangement

Figure 6.7

are installed, sealed, and the main reactivated. During the installation of this connection, the water main is out of service for a considerable length of time, and all the suppression systems and other utilities the main serves are without water and, therefore, out of service.

The second method, the wet tap, consists of a connection to an active water main without interruption of the water supply. A wet tap installation begins with a tapping sleeve bolted tightly around the pipe at the point of connection. The tapping sleeve has an outlet the size of the branch connection to the main. A non-rising stem valve, or *tapping valve* is bolted to the outlet on the tapping sleeve, the valve is opened, and the tapping machine bolted to the valve. The tapping machine extends through the valve and cuts the opening in the water main to the exact size required for the connection. The cutter is removed along with the pipe segments, the tapping valve closed, and the tapping machine removed. (See Figure 6.8.)

Application of Tapping Machine to Pipe under Pressure

Figure 6.8

Before the tapping machine begins cutting the water main, a small pilot drill is inserted in the center of the piece of pipe to be cut out. This drill will retain the cut piece of pipe so it cannot drop into the water main, and assists in the removal of the tapping machine. Once the wet tap is completed, the new underground branch can be connected to the closed tapping valve outlet, the valve can be opened, and the new main filled with water.

As previously mentioned, with a wet tap there is no interruption to service when making a connection to an existing active water main. Municipalities usually request or require a wet tap so as not to interrupt domestic or fire protection service. In addition, the tapping valve becomes a key city valve, giving them the ability to shut off the water service should this ever become necessary. The wet tap is also used when it becomes necessary to make a connection to existing fire mains on private property.

Strainers

When a system has sprinkler heads with orifices of less than 3/8" or a water spray system where the orifices can be very small, a strainer must be installed in the main supply line or the system riser. Another need for a strainer arises when the water supply is from a source where foreign material could enter the piping. Strainers for fire protection use are UL listed and/or FM approved, and contain a strainer and a blow off connection for flushing.

Hangers

Hangers that are used to support fire suppression system piping from the building structure consist of two or more UL listed components. The *hanger fastener* and the *hanger pipe ring* must both be UL listed components. In addition, if the hanger assembly has a rod between the fastener and the pipe ring, it must be sized according to the NFPA requirements. The hangers must support the piping when it is filled with water, which, according to NFPA must be the weight of the water-filled pipe plus 250 pounds, applied to the point of support.

The hangers used to support fire protection piping cannot support any other piping, and must be attached directly to the building structure. If piping is run under large ducts, the hangers are attached either to the building structure or to the angles that support the ducts, provided angles meet the size requirements of the NFPA. When fire protection system piping must be supported by any nonstandard assembly, it must be certified by a professional engineer. This certification will verify that the hanger installation is capable of supporting five times the weight of the water-filled pipe plus 250 pounds at each hanger. Detailed calculations must be submitted to the NFPA by the certifying engineer indicating all the potential stresses and safety factors. With the multitude of UL listed hanger components and assemblies, a special hanger is seldom required.

The location of fire protection system hangers is detailed in the NFPA Standards, especially NFPA Standard 13 for sprinkler systems, and these requirements must be carefully followed. The important aspect of this brief section on hangers for fire protection systems is to impart the serious necessity of:

1. Using only UL listed hanger components
2. Installing hangers according to the requirements of NFPA
3. Installing hangers in locations on the piping where required by NFPA

If the fire protection systems are designed and specified by individuals not familiar with NFPA requirements and the use of UL listed hanger components, problems will ensue. For example:

A fire protection system may be designed and specified—by a power plant designer not familiar with fire protection support requirements. They use specific methods and hangers in their work, and if they are not acquainted with the specific requirements for hanging fire protection piping and the use of UL listed hanger components, they may specify pipe hangers which are not rated. When the bidders get the design and specifications, they know they cannot bid the job using the hangers specified, and must take exception to the bid request. They would request an addendum to clarify the situation and put all bidders on an equal footing.

CHAPTER SEVEN

UNDERGROUND FIRE MAINS

CHAPTER SEVEN

UNDERGROUND FIRE MAINS

Selection of underground pipe for fire protection depends on several considerations: an evaluation of the soil conditions, the maximum working pressure, and the external loads the pipe may experience by virtue of vehicular and other traffic. Fire protection water mains must be capable of withstanding the unusual, and sometimes severe, conditions associated with fire protection systems:

- fire pump pressure surges
- hose stream operation
- deluge system operation
- the sudden closing or opening of valves

Cast iron, ductile iron, steel, and asbestos cement underground piping do not carry either UL listing or FM approval. This pipe, when used for underground fire mains, is acceptable to Factory Mutual and to the insurance carriers who normally rely on UL listed equipment, as long it is manufactured according to the standards of the American Water Works Association (AWWA) and the specifications of other nationally recognized engineering organizations. This chapter explains underground pipe selection, hydrant selection, flushing, and testing for the various types of piping listed above.

Underground Pipe Selection

Suppression systems that use water as their extinguishing agent are usually designed for a maximum working pressure of 175 psi. If the system pressure is expected to exceed 175 psi, extra heavy fittings (250 lbs., 300 lbs.), valves, and devices must be used. In addition, the sprinkler heads must be UL listed, another factor that adds to the system installation cost. Systems that will not exceed this pressure limit can use standard weight fittings, valves, and sprinklers.

System pressure levels also have a design impact on both the underground fire mains and the above ground system. For example, pump installation requires considerable experience to balance pump shut off pressure of 120% of the rated head for vertical turbine pump suction and system demand, while maintaining the system working pressure below 175 psi. Shut off pressure can be particularly difficult while attempting to size a fire pump to deliver the system demand without exceeding the 175 psi level. As an example: In a system with a demand pressure of

150 psi, a horizontal split case fire pump must be used, which will deliver 180 psi at shut off pressure and require the use of extra heavy fittings and valves.

Cast Iron Pipe

For standard systems (working pressure below 175 psi) underground cast iron and ductile iron pipe rated as Class 150 is most commonly specified. The Class 150 rating represents the main working pressure. This distinction can be confusing, as compared with the 175 number. The 150 psi classification for cast iron and ductile iron pipe (and most flanged valves) is the maximum working pressure with *high temperature fluids*. At the normal operating temperatures of suppression systems, namely transporting water, the 150 psi working pressure is adequate to handle pressures in excess of the 175 psi. These ratings are normally cast into the valve bodies: For example, 125 WSP (working steam pressure), and 200 WOG (water, oil, gas).

For 175 psi suppression systems and the laying conditions previously described (namely 5′ bury and a tamped backfill) the pipe should be listed as "Class 22, 150 pound working pressure."

If system pressure will exceed 175 psi, cast iron underground pipe Class 200, 250 and up should be considered based on an evaluation of all of the conditions involved.

Ductile Iron Pipe

This type of piping is, as its name implies, much more flexible than cast iron. Ductile iron will bend considerably before it will ultimately fail and exhibits a tremendous impact strength. One manufacturer of ductile pipe demonstrated its strength by dropping a length of pipe from a great height onto a rail without having the pipe length break. In another demonstration, a 4,500 pound weight was dropped on a length of ductile iron pipe, and the pipe was flattened but did not break.

The strength of ductile iron pipe makes it desirable for use under heavy vehicular traffic conditions and in unstable soil where the pipe could be subjected to the stress of traffic loads. The tremendous pressure impact of a water hammer has no effect on ductile iron pipe, where cast iron might crack or rupture.

Specifics for Cast Iron and Ductile Pipe

Cast iron and ductile iron pipe for fire protection should be lined pipe (cement/enamel/polylined), to prevent the buildup of corrosion within the pipe interior which would increase the resistance to water flow in the pipe. After being installed for several years in a highly corrosive water condition, this corrosion buildup can greatly reduce the internal diameter of the pipe. Cast iron and ductile iron pipe joints can be bell and spigot, mechanical, and push-on type.

The spigot, or plain end of the pipe, is inserted into the bell end of the pipe; packing, usually jute, is fitted into the bell around the spigot end; and the joint is sealed with molten lead.

Although bell and spigot use to be the most common type of cast iron pipe joint, they require considerable labor to assemble. As a result, bell and spigot joints are seldom used in fire protection mains in today's labor conscious market.

Mechanical joint pipe (see Figure 7.1) is basically flanged piping bolted together, with rubber gaskets sealing the flanged joints.

Push-on joints are basically a modified bell and spigot arrangement where the plain end of the pipe is inserted—pushed—into the bell end over a rubber or neoprene O-ring type gasket, compressing it, thereby sealing the joint.

Friction Loss — The "C" Factor

The coefficient, or constant, for friction loss in underground piping is represented by the letter "C," followed by an indicating number. The larger the number following the letter "C," the less friction loss in the pipe. For example:

Cast iron—unlined—new   C = 120
Cast iron—lined—new   C = 140

Flanged, Mechanical, and Push-On Joints

Figure 7.1

Cast iron in use (public or industrial main) for moderately corrosive water:

10 years old	C = 90
20 years old	C = 65
30 years old	C = 55

These "C" factors can make a major difference in pressure loss when hydraulically calculating the underground piping supply to fire protection systems. For example:

1,000 gpm, 6" cast iron pipe, C = 100,
  friction loss per 1,000' of pipe is 52 psi.

1,000 gpm, 6" cast iron pipe, C = 140,
  friction loss per 1,000' of pipe is 27.92 psi.

Other Types of Underground Piping

There are some instances in which soil and/or water conditions can cause considerable corrosion with cast iron or ductile iron pipe. The water in the soil may be extremely corrosive to these pipe materials, because of the acidity and alkalinity of the area. Through a complicated electrochemical reaction, external corrosion of cast iron and steel pipe can be relatively rapid. Stray currents can follow the buried piping to a point where the soil has less electrical resistance than the pipe, and the current leaves the pipe. This creates ionization that is similar to soil corrosion. When a soil survey reveals these electrical currents from an external source, there are two methods of preventing this current from causing severe erosion of the cast iron pipe—namely *coating and wrapping the pipe* or using *cathodic protection.*

Currents can be bled-off by providing low resistance metallic grounding connections. Basically, cathodic protection applies a direct electric current to the pipe line to nullify the stray current.

Polyvinyl Chloride Plastic Pipe (PVC)

A reliable substitute for cast or ductile iron pipe, polyvinyl chloride (PVC) piping is highly resistant to corrosion, is available in Class 150, and is lighter in weight. In fact, its relatively light weight provides a considerable labor savings cost impact due to ease of handling and installation. This comparison between the additional material cost of PVC over the material cost of handling and installation and the decreased cost of installation for PVC over cast iron, must be carefully evaluated if cost is the primary reason for using PVC pipe.

PVC can be field cut with a power saw, and uses a modified bell and spigot type joint with a ring placed in a groove in the bell end that provides a tight seal when the spigot end is installed in the bell end. Flanges can be attached to the PVC to accommodate attachment to cast iron valves or fittings. Being nonmetallic pipe, PVC is not subject to external or internal corrosion, nor will it experience the effect of electrical currents that cause external corrosion.

The "C" factor for PVC is 150, compared with new lined cast iron and ductile iron pipe with a "C" factor of 140. PVC underground pipe may be required due to soil conditions that would corrode other piping materials, or to take advantage of the improved "C" factor of 150. If there is a considerable amount of underground fire main involved, the use of PVC could mean the difference in hydraulically calculating the demand between a large cast iron or ductile iron pipe size, and a smaller PVC fire main. (See Chapter 8 for an explanation of hydraulic calculations.)

Fiberglass-reinforced Plastic Pipe

Another nonmetallic underground pipe which resists corrosion, fiberglass-reinforced pipe also does not require cathodic protection. Like PVC, this pipe is lighter than cast or ductile iron, and the joints are of a modified bell and spigot type, where the spigot end is coated with an adhesive before being inserted into the bell end.

Asbestos Cement Pipe

This pipe is another material that does not corrode, and usually consists of a sleeve (coupling) with an O-ring gasket at each end. The ends of the pipe are tapered, and when inserted into the sleeve they compress the gaskets which make a tight joint. The "C" factor for asbestos cement pipe is 140.

Steel Pipe

Steel pipe can be used for buried fire mains using welded joints with the pipe coated and wrapped, but even with this corrosion precaution, the corrosion factor of buried steel pipe is a major concern. If it ever becomes necessary to bury a dry pipe system, the use of steel pipe with welded fittings and joints is the only means of retaining the compressed air without excessive loss.

Fire mains smaller than 6" may be used, but they must be lined, their use must be proven by hydraulic calculations, and the main must be as large as the system riser. The cost of installing a 6" fire main is almost equal to installing a 4" main. The material difference is very minor, and the trench work almost identical, so always consider the advantages of a 6" supply to that of a 4" in relation to the slight cost increase.

Underground Pipe Bury

The *bury*, or cover of earth over the top of the pipe, depends on the temperature of the area (which determines the depth of the frost line). Maps are available which indicate the recommended bury depth of piping throughout the country determined by the maximum depth of frost penetration. In locales where frost is not a consideration, underground piping should not have less than 2-1/2' of bury to protect the piping from mechanical damage. Where freezing is a factor, depth of static or stationary fire protection piping should be 6" deeper than municipal water mains which have a regular flow of water.

The term "bury" is used to establish the cover on the pipe. When determining the required depth of the pipe trench, be certain to provide the required bury while compensating for site grading.

When underground piping passes under a building foundation, the bury should be the same as that required for outside bury—a sleeve through the foundation wall is the most common solution. There are times when it is absolutely necessary to run the fire main under a footing that is at greater depth than the pipe line. When this occurs, the pipe must not be raised to pass above the footing, otherwise it will not have the required bury.

Extreme caution must be taken in fill compaction in the area to prevent future settling which can shear through the pipe line.

This condition mandates other innovative solutions, which may prove to be more expensive; for example, trenching at a greater depth for the entire run of piping or placing 45 degree ells to drop the line under the footing. However, violating the required bury leaves open the potential of the water laying static in the pipe freezing, and subsequently cracking the pipe, which is much more expensive than the innovative installation solution. Even if the freezing is not severe enough to crack the pipe, any ice formation will obstruct the water flow that the suppression system needs in a fire.

Rodding and Thrust Blocks

Tremendous thrusts are created in the buried piping wherever the pipe changes direction, due to pressure surges in fire protection underground piping when flow is started or stopped. Most cast iron and ductile iron pipe, bell and spigot, or mechanical joints, will use rods that run from fitting to fitting, or fitting to joint to hold the joints, fittings, and valves in opposition to these thrusts and relieve the pressure that could cause a joint to separate.

When soil conditions are not stable enough to hold a fitting in place when a water surge occurs, it is necessary to install thrust blocks. Under any number of conditions—excessive pressure in the main, unstable soil—thrust blocks may be required, even if the piping is rodded. This of course requires a thorough engineering evaluation.

Thrust blocks are masses of poured concrete, poured against the fitting where the pipe changes direction, or against a valve that could be subject to surge pressure. (See Figure 7.2.) The consistency of the concrete, the size of the thrust block, and the technique for pouring the thrust block must be in accordance with the authority having jurisdiction. It is also this authority, especially if they are an insurance carrier, who may dictate the use of thrust blocks, rodding, and retainer glands (devices used on mechanical joint pipe to prevent separation of joints). Whether or not the stability of the soil is known, it is always advisable to meet the insurance requirements.

Bell and spigot pipe can be rodded by installing a pipe clamp and rodding through the pipe clamps. But in the majority of cases there is no question that nonmetallic pipe like PVC must have thrust blocks. There are very few instances where the use of thrust blocks will not be required when using piping other than mechanical joint cast iron or ductile iron piping.

Other Considerations

Underground piping must never be installed below the floor of a building. This is against NFPA Fire Codes and good engineering practice. Installing piping below the floor means that a leak or break would not be detectable, and could undermine the floor, causing collapse. Even if the leak or break were detected through a meter, finding the location of the leak and repair would be a costly operation.

When there is an underground piping system that supplies several hydrants and/or several suppression systems, or even several buildings in an industrial or office complex, the fire main should be *sectionalized*. This means installing post indicator valves (PIV's) at various locations so that, should a repair of the fire main be required, the entire fire main supply would not have to be shut down—the section between two sectional PIV's could be isolated by closing just these two valves.

There is a rule of thumb that *sectional valves should not control more than five systems*, with hydrants being considered as a "system." This rule exists because the placement of sectional valves must be reviewed and approved by the insurance interest having jurisdiction.

Hydrants   The installation of fire hydrants supplied by a private fire main involves three types of hydrants. One is a *wet barrel hydrant* used in areas that are not subject to freezing because the water fills the hydrant exposed above ground. The second type of hydrant is a *dry barrel hydrant* (See Figure 7.3) where the water is kept at its base by a valve that is opened when the hydrant nut is operated. The third type is the *wall hydrant* that has been previously described. It is installed on the wall of the building, is similar in appearance to the fire department siamese connection, and water is controlled by a non-rising stem valve located inside the building with the operating nut on the exterior wall.

In most cases, the hydrants that are installed on private property supplied by private fire mains consist of 2-1/2" hose outlets. Most public hydrants also have a larger, usually 4", pumper connection, but insurance carriers will not permit the pumper connection on private hydrants. They are usually concerned that too much water is being taken from this outlet, thereby decreasing the supply to the suppression systems. To cite an actual project experience:

> The hydrants for installation on a private property fire main were incorrectly ordered with the 4" pumper connection. The insurance carrier required the owner to weld the 4" outlets shut so there was no chance they would be used.

Private hydrants should be located approximately 300' to 500' apart, using the 300' spacing in areas of greatest hazard, and 500' between hydrants in other areas, but no hydrant hose length should exceed 500'. Placement of hydrants becomes an experienced engineering judgment, and here again, their location is subject to the authority having jurisdiction.

Thrust Block

Figure 7.2

Flushing and Testing

Above-ground suppression systems must be hydraulically tested, but underground fire mains must be flushed and tested. The piping must be flushed at the water demand rate of the system, and this demand flow must be at least of a capacity to produce a velocity in the flushing water of 10' per second. To produce this velocity, the following gpm must be discharged through the appropriate pipe size:

4" pipe	390 gpm
6" pipe	880 gpm
8" pipe	1560 gpm
10" pipe	2440 gpm
12" pipe	3520 gpm

After the piping is installed, and before conducting the hydrostatic test, the installing contractor should not backfill and cover the piping joints or valve connections. However, it is mandatory to backfill the trench between the joints to prevent any pipeline movement during the pressure test.

Dry Barrel Fire Hydrant

Figure 7.3

196

Underground piping must be tested at 200 psi for 2 hours. When the static pressure exceeds 150 psi, the piping must be tested at 50 psi over the maximum static pressure. During the hydrostatic test, determine the amount of leakage by pumping water into the system from a container that has been calibrated. New underground piping installations are not permitted a leak more than two quarts per hour per 100 gaskets or joints. There are also minor leakage allowances for hydrants supplied from the fire main that are tested at the same time as the fire main.

All flushing and tests must be witnessed by the authorities having jurisdiction, the owner, or appointed representatives, and at the conclusion of the flushing and tests, the installing contractor must furnish a signed "Contractor's Material and Test Certificate," countersigned by the witnesses. If the installing contractor is not performing the tests, the above ground system contractor must receive a copy of the flushing and test certificate. The above ground contractor is assured that the main has been flushed and tested, and any problems with obstructions in the above ground system are his responsibility.

To complete the testing of the underground piping system, each hydrant supplied by the fire main must be opened and closed while the system is under pressure to uncover any hydrant leaks. In addition, all control valves on the underground fire main must be closed and opened while the system is under operating pressure to ensure that the valves are opening and closing properly.

CHAPTER EIGHT

HYDRAULIC CALCULATIONS

CHAPTER EIGHT

HYDRAULIC CALCULATIONS

There are two systems for establishing the size of system piping: Pipe schedule and hydraulic calculations. The pipe schedule method uses tables that indicate that steel pipe of a certain size will supply a certain number of heads (see Figure 8.1). Automatic sprinkler systems are classified as light hazard, ordinary hazard, and extra hazard based on the building use or type of occupancy the system it is designed to protect, and there are different pipe size schedules for each of these three classifications.

Pipe Schedule

While many pipe installations have been performed by the pipe scheduling method, hydraulically calculated systems end up being more efficient and economical. However, an explanation of the pipe schedule method is presented here for comparison. Pipe schedule sizing does not apply to systems using other than 1/2" orifice heads, and cannot be used when the system is to be hydraulically calculated.

The reference to maximum floor area allowed to be supplied by one system or one riser for the given pipe size means that if a one-story building has 52,000 square feet for light hazard or ordinary hazard occupancies, or 25,000 square feet for extra hazard occupancies, the entire area can be supplied by the 4" system for light hazard, 8" for ordinary or extra hazard.

If the building is multi-story with floors separated by complete floors with no large openings, each floor can be supplied by a riser sized to protect the maximum floor area given in the pipe schedule. (See Figure 8.1.)

If the floors separating the different floors have large openings that cannot be sealed or closed off, the risers supplying the sprinklers must be sized for all the heads on all the floors. A fire on one floor could open the heads on the fire floor and then spread up through the large openings and open heads on the next floor, so the riser has to be sized to supply the heads on the fire floor and all other floors where heads may open.

Open head deluge systems cannot have the pipe sized by pipe schedule because all the heads will be discharging when the deluge valve trips.

If branch lines exceed eight heads, the two end pieces of pipe should both be sized 1-1/4"; if the branch lines have 10 heads, the two end pieces must be 1-1/4" and the pipe feeding the 10 sprinklers must be increased to 2-1/2". This rule applies to light and ordinary hazard occupancies.

LIGHT HAZARD PIPE SCHEDULE

Steel Pipe

1	inch pipe	2 sprinklers
1-1/4	inch pipe	3 sprinklers
1-1/2	inch pipe	5 sprinklers
2	inch pipe	10 sprinklers
2-1/2	inch pipe	30 sprinklers
3	inch pipe	60 sprinklers
3-1/2	inch pipe	100 sprinklers
4	inch pipe	The maximum floor area protected by sprinklers that are supplied by one system or one riser. (52,000 sq. ft.)

ORDINARY HAZARD PIPE SCHEDULE

Steel Pipe

1	inch pipe	2 sprinklers
1-1/4	inch pipe	3 sprinklers
1-1/2	inch pipe	5 sprinklers
2	inch pipe	10 sprinklers
2-1/2	inch pipe	20 sprinklers
3	inch pipe	40 sprinklers
3-1/2	inch pipe	65 sprinklers
4	inch pipe	100 sprinklers
5	inch pipe	160 sprinklers
6	inch pipe	275 sprinklers
8	inch pipe	The maximum floor area protected by sprinklers that are supplied by one system or riser. (52,000 sq. ft.)

EXTRA HAZARD PIPE SCHEDULE

Steel Pipe

1	inch pipe	1 sprinkler
1-1/4	inch pipe	2 sprinklers
1-1/2	inch pipe	5 sprinklers
2	inch pipe	8 sprinklers
2-1/2	inch pipe	15 sprinklers
3	inch pipe	27 sprinklers
3-1/2	inch pipe	40 sprinklers
4	inch pipe	55 sprinklers
5	inch pipe	90 sprinklers
6	inch pipe	150 sprinklers
8	inch pipe	The maximum floor area protected by sprinklers that are supplied by one system or riser. (25,000 sq. ft.)

Figure 8.1

With extra hazard pipe schedule systems, the branch lines must not exceed six sprinklers on either side of the cross main.

With light hazard pipe schedule systems, if the area being protected does not have walls to partition off the area (uncompartmented), and the number of heads exceeds 100, the sprinklers must be supplied by pipe sizes from the ordinary hazard pipe schedule.

When the spacing between sprinkler heads exceeds 12', or the distance between the sprinkler lines exceeds 12', only 15 sprinklers can be supplied by 2-1/2" pipe, 30 sprinklers supplied by 3" pipe, and 60 sprinklers supplied by 3-1/2" pipe, for ordinary hazard pipe schedule systems. The remaining heads can be supplied by the pipe schedule for ordinary hazard systems for 4", 5", 6", and 8" pipe sizes.

Hydraulic Calculations

The design criteria used in hydraulic calculations is also based on three classifications, but ordinary hazard classification is subdivided into Ordinary Hazard Group 1, Ordinary Hazard Group 2, and Ordinary Hazard Group 3. Extra hazard is also divided into Extra Hazard Group 1 and Extra Hazard Group 2 (See Figure 8.2).

When using the pipe schedule to size sprinkler piping, a table in NFPA Standard 13 provides the requirements for water supply. This table is broken down into requirements for water supply, and requirements for light and ordinary hazard, groups 1 and 2. Ordinary Hazard Group 3 and extra hazard requirements for water supply must be approved by the insurance carrier.

When using hydraulic calculations to size the system piping, another table in NFPA Standard 13 can be used to obtain the design criteria for the different classifications, or the design criteria will be provided by the insurance carrier. Since the piping schedule method is well known, this chapter addresses the hydraulic calculation method for sizing pipe. The purpose of this chapter is not to instruct the reader in how to make hydraulic calculations, but to present the concepts so as to provide familiarity with the procedure and take out some of the "mystery."

Design Criteria

By hydraulically calculating the area operation, one is saying that the pipe is sized to deliver the required flow and pressure each head in the area of operation and sizing the pipe back to the system water supply in order to deliver the demand for the area of operation.

Hydraulic calculations use design criteria which is based on the type of occupancy, and mandates that each head on the system will discharge not less than a specified, or predetermined, number of gallons per minute per square foot of the protected area. For closed head sprinkler systems, there must also be an area of operation, or area of application, which is the maximum square footage a fire in that type of occupancy has the potential to cover. This figure establishes the quantity of sprinklers required open to contain and extinguish the fire within the area of operation. This area also establishes the number of heads to be hydraulically calculated:

> Divide the square footage of the area of operation by the square footage each sprinkler head covers. If each sprinkler head covers 100 square feet, and the area of operation is 3,000 square feet, 30 heads will have to be calculated.

Over this area of operation, every sprinkler head must have sufficient volume and pressure to discharge at least the predetermined and stipulated gallons per minute per square foot. This area of operation must be the area that is furthermost removed, hydraulically, from the source of supply. This is commonly referred to as *the most hydraulically remote square feet.*

If the gpm per square foot can be accomplished by hydraulic calculations in the most remote area hydraulically from the source of system water supply, all the heads in the other areas of operation (closer to the system water supply) will discharge the required gpm per square foot, or more,

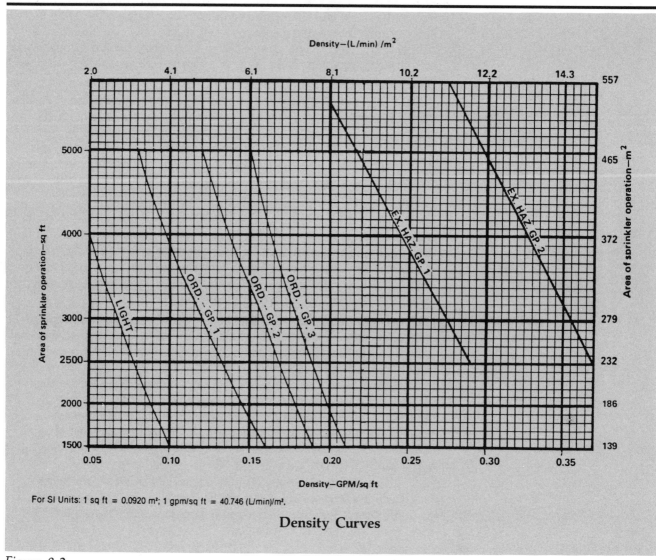

For SI Units: 1 sq ft = 0.0920 m²; 1 gpm/sq ft = 40.746 (L/min)/m².

Density Curves

Figure 8.2

since the pipe friction loss to reach these closer areas of operation is less than that required to reach the remote area of operation. For example:

> The design criteria for this facility is .30 gpm per square foot over the most hydraulically remote 3,000 square feet. Therefore, it is only necessary to determine which 3,000 square feet are the most remote, hydraulically, from the system water supply. With each head covering 100 square feet, hydraulically calculate the 30 heads in this area. As mentioned previously, all the other areas of operation (the 3,000 square feet) closer to the supply will receive the same, if not more, than the water demand in the remote area.

As stated, hydraulically calculating the heads in the remote area of operation means *sizing the piping* to these heads so that each head will be discharging not less than the design criteria mandates, namely .30 gpm per square foot or in the example with 100 square feet per head, 30 gpm per head (.30 times the 100 square feet).

When calculating a deluge system, the design criteria will state so many gpm per square foot over the entire area, since every head on a deluge system is the entire area covered by the system.

Flow Data

In order to hydraulically calculate a system, the flow data must be available. *Flow data is the supply criteria.* This is the static pressure, or the pressure in the supply main with no water flowing. In addition, the residual pressure is that left in the water supply main when water is flowing. But residual pressure is of no value unless the amount (gpm) of water flowing at the time residual pressure is obtained is known. It is imperative to obtain all three facts:

- static pressure
- residual pressure
- flow in gallons per minute when the residual pressure was read on the gauge.

All three of these items are necessary to prepare a supply graph. Figure 8.3 indicates the supply graph for a water supply from a source like a city water main—it is a straight line. When the system demand is calculated, it is plotted on this graph. If the demand is below the supply line, the supply will satisfy the demand, but if the demand is above the supply line, the supply will not satisfy the demand. It is always recommended and should always be specified, and is sometimes required by the authority having jurisdiction, to have the calculated system demand fall at least ten psi below the supply line. This is important because flow data on a water main can depend on peak usage hours. For this reason, the information sheet on a set of hydraulic calculations that describes the flow data must include date, location, day of the week, and even time of day.

Flow data obtained from a flow test on a city main performed at ten o'clock on a Sunday would obviously produce much better results than if the same flow data were obtained at 10 o'clock on a Monday when domestic and industrial use of water was at a much greater level. This example is an exaggeration, but it illustrates that flow data can change during the course of a day or week, and the 10 psi safety factor of the demand below the supply curve allows for these periodic "normal" changes in the flow data.

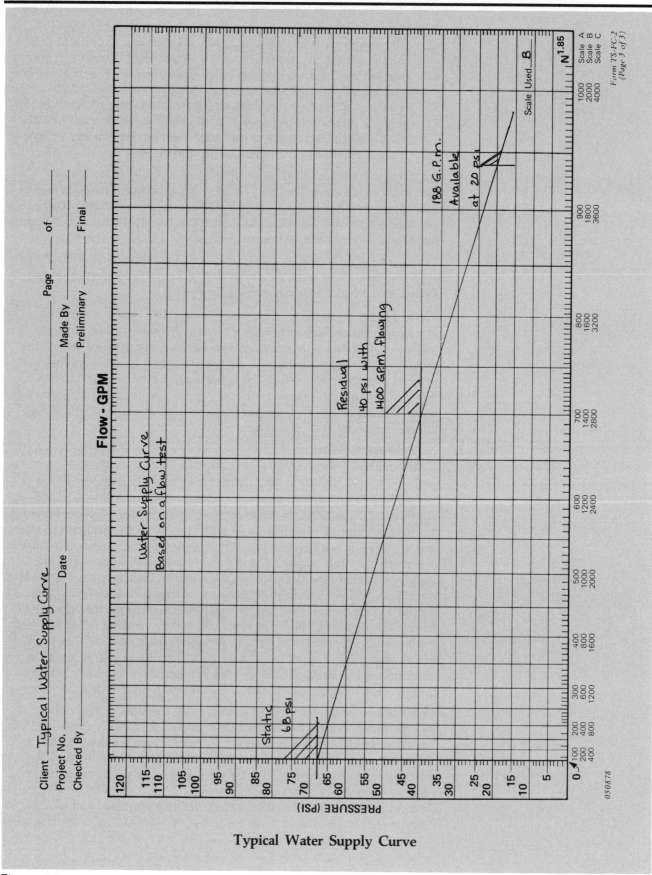

Typical Water Supply Curve

Figure 8.3

Keep in mind that the supply curve for a fire pump is actually the pump curve, or a combination of the two, which reflects what residual pressure the pump suction is receiving when it is delivering the required demand volume.

Flow Test

Flow data from a municipal or private water supply is obtained by conducting a flow test. A flow test consists of placing a pressure gauge on one 2-1/2" outlet of a hydrant and opening the hydrant nut to admit water. No water is flowing because the pressure gauge is screwed into the hydrant outlet cap, and the other caps on the hydrant are still in place. With no water flowing from this hydrant (it is now under pressure), the static pressure is indicated on the gauge.

One person is stationed at this hydrant with the pressure gauge, and another goes to a second hydrant on the water main, and, with the hydrant still closed, removes a 2-1/2" hydrant cap. When both individuals are "ready," the second person opens the hydrant to allow water to flow from the open 2-1/2" hydrant butt, or outlet. When this individual notices that the flow from the hydrant is constant, a *pitot tube* is inserted in the flow. This tube measures the velocity pressure of the flowing water, which is recorded on a pressure gauge attached to the pitot tube. When the person feels that the reading on the tube pressure gauge has stabilized, a signal is given to the other person at the static pressure gauge hydrant, and the static pressure gauge, which is now registering residual pressure, is recorded (see Figure 8.4).

When the test is over and the hydrants returned to their normal condition, the reading on the tube pressure gauge is compared with a chart which converts velocity pressure to flow in gallons per minute. The flow data is complete: static pressure, residual pressure, and flow in gpm at the residual pressure.

This is a simplified explanation of a flow test. There are certain items that must be taken into consideration, such as the size and shape of the hydrant butt. These factors influence the velocity pressure of the hydrant discharge, and the chart that converts velocity pressure to flow has allowances for these conditions in the form of different columns on the chart, or a factor to multiply the velocity pressure by to obtain a true reading.

Approximate Calculated Demand

The approximate gpm demand of a system can be obtained from the design criteria by multiplying the density, gpm per square foot, by the area of operation.

To bring this figure up to a more accurate demand figure and closer to what will be the gpm at the bottom of the hydraulic calculation sheet, multiply the answer by 1.15. The friction loss in the pipe in the calculations will increase the psi at each head and subsequently result in slightly more gpm at each head. The end head in the area of operation where the calculations begin is the only head that will discharge exactly (in our example) 30 gpm—.30 times the area covered by head, namely 100 square feet.

In order to discharge 30 gpm, this head requires a specific amount of pressure, in this case approximately 29 psi. In getting 30 gpm through the 1" pipe 10' long there will be almost a three psi loss due to friction in this length of pipe. Therefore, at the second head, the psi will have to be

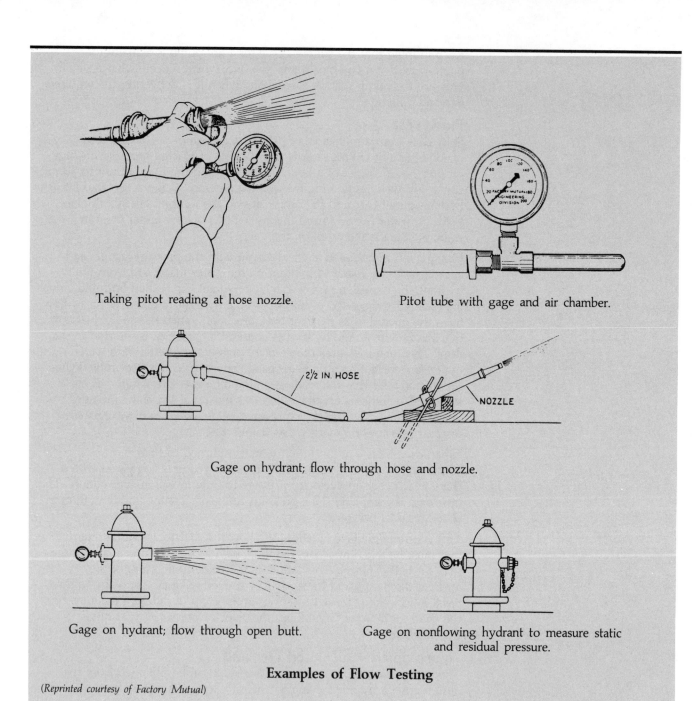

Taking pitot reading at hose nozzle.

Pitot tube with gage and air chamber.

2½ IN. HOSE

NOZZLE

Gage on hydrant; flow through hose and nozzle.

Gage on hydrant; flow through open butt.

Gage on nonflowing hydrant to measure static and residual pressure.

Examples of Flow Testing

(Reprinted courtesy of Factory Mutual)

Figure 8.4

29 psi plus three psi, or 32 psi to deliver 30 gpm to the end head at 29 psi. This means that the second head will be discharging at 32 psi, approximately 32 gpm. This increase in pressure at each succeeding head increases its discharge.

The example above is extreme, since the friction loss of approximately three psi is in all probability too great, and the pipe size would have been 1-1/4" to reduce the friction loss to .74 psi in lieu of 2.76 psi with the 1" pipe. This does point out the necessity to adjust pipe sizes in performing hydraulic calculations in order to maintain the demand, both in volume and pressure, remain below the supply curve.

The foregoing illustrates the reason to increase the design criteria by 15% and even 20% to obtain a realistic demand figure. As the stamp on the preliminary plan states, this figure is preliminary and not for construction. The only figure that can be used to size fire pumps, storage volumes underground mains, etc., is the figure at the bottom of the hydraulic calculation sheets.

Computer vs. Hand Calculations

Most hydraulic calculations for fire suppression systems are currently performed by computers. The extremely simplified example presented later in this chapter points out one minor reason for performing hydraulic calculations by use of a computer program. Had this calculation been performed by hand, and the excessive friction loss been reflected in too small a pipe sizing throughout the calculation, the demand would probably have been above the supply curve, the calculation sheets thrown away, and the calculations redone with adjusted pipe sizing to reduce friction loss. By computer, these adjustments could be made and the calculations redone almost at the flick of a few keys.

The hydraulic calculations prepared by hand that have been included in this text are merely presented to show the steps involved in preparing hydraulic calculations. Including these calculation sheets is basically an effort to establish the procedure for calculating a sprinkler system by "walking the reader through" a sprinkler calculation to offer the basic concept of hydraulic calculations. The hand calculations are based on a "Gridded" sprinkler layout: One common supply main (the trunk) feeds the sprinkler lines (the branches). (See Figure 8.5.)

Sprinkler systems where the lines are all joined together are known as *gridded systems*. These systems allow much smaller pipe sizes, but are next to impossible to hydraulically calculate by hand. The calculations must be done by computer (unless performed by a highly experienced engineer).

Necessity of Hydraulic Calculations

Unless the system is laid out, hydraulically calculated, and the pipe sizes derived from the calculations indicated on the plans, a detailed cost estimate cannot be prepared. If the design criteria is given but a preliminary system design with pipe sizes is not presented, the estimator will not have any concept of pipe sizing.

Many performance specifications will provide the design criteria, but because it is a performance spec, no preliminary system design will be offered, and the estimator is at a loss for pipe sizes and lengths.

Caution: If the engineer on a project elects to use his or her experience and judgment in selecting the design criteria for the hydraulic calculations from the NFPA Standard 13 table, it is always advisable to verify the design criteria with the insurance carrier. Insurance carriers know the operation and hazards of their clients, and if the engineer selects a design criteria that is not adequate to satisfy the insurance requirements, this can cause additional and unnecessary expense in recalculating, redesign, and possibly retrofitting the installation.

Summary Hydraulically calculated systems provide assurance that an efficient and effective amount of water will be discharged on the combustion. With a pipe selection system, it can only be assumed that the combustion will be suppressed and/or extinguished by the system discharge.

There is an industry adage that if you give five sprinkler designers a building to be protected, you will get five different system designs. The same holds true of hydraulic calculations and the resultant pipe sizing.

The previously mentioned performance specification is a variation in layouts and calculations. The performance system and specification leaves the design up to the bidder or contractor. Usually only the point where the system water supply enters the building is indicated, and the design, system criteria, and parameters are covered in the specifications. This can sometimes lead to a wide variance in bids because one bidder may have figured a gridded system and another a "Christmas tree" system.

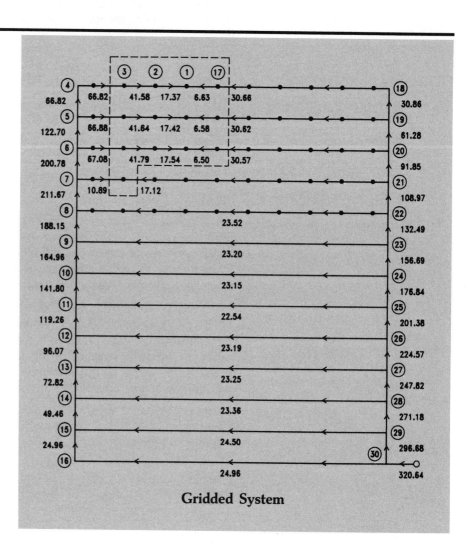

Gridded System

Figure 8.5

In addition, the architect and engineer generally have little control over the system layout. When shop drawings (installation plans) are received to be checked, the architect/engineer has little recourse as long as the system design meets specifications and satisfies design criteria. The architect/engineer may recommend design alterations if the proposed layout does not meet expectations relative to appearance or potential maintenance or space allocation problems.

A preliminary system design with pipe sizes determined by hydraulic calculations will correct some of these problems, and all the bids will be based on the same factors, but the preliminary design and pipe sizes may not be the most economical or efficient. It is, therefore, always recommended that the specifications allow the successful bidder to redesign and recalculate the system based on the specification requirements, subject to the approval of the engineer.

For the different types and characteristics of pipes used in fire protection systems, refer to the Appendix.

The next portion of this chapter contains Figure 8.6, Hydraulic Calculation Sheets, and an explanation (Figure 8.7) of how these figures were derived.

Client _Curfew Company_   Page _1_ of _4_

Project No. _13903.300_   Date _12/90_   Made By _W.C._

       Checked By _A.N._

       Preliminary _____ Final ✓

Hydraulic Design Information Sheet

Name _Curfew Company_   Date _12/12/90_

Location _Kansas City, Missouri_

Building _Industrial - Machine Shop_   System No. _1_

Contractor _B + V Sprinkler Co._   Project No. _13903.300_

Calculated By _Walter Coon_   Drawing No. _1 of 1_

Construction: ☐ Combustible   ☒ Non-Combustible   Ceiling Height _25_ Ft.

Occupancy _Machine shop with hydraulic presses_

System Design

☒ NFPA 13:   ☐ Lt. Haz.   Ord. Haz. Gp.   ☐ 1   ☐ 2   ☒ 3   ☐ Ex. Haz.

☐ NFPA 231   ☐ NFPA 231C: Figure _____ ; Curve _____

☐ Other (Specify) _____

☐ Specific Ruling _____   Made By _____   Date _____

Area of Sprinkler Operation _2800 S.F._	**System Type**
Density _0.25 GPM/SF_	☒ Wet ☐ Dry ☐ Deluge ☐ Pre-Action
Area Per Sprinkler _100 S.F._	**Sprinkler or Nozzle**
Hose Allowance GPM: Inside ⎰ _650 GPM_	Make _Viking_ Model _S.S.U._
Hose Allowance GPM: Outside ⎱	Size _17/32 - inch_ K-Factor _8.2_
Rack Sprinkler Allowance _Not Applicable_	Temperature Rating _165°F_

Calculation Summary   GPM Required _848_   PSI Required _86_   At Base of Riser

"C" Factor Used: Overhead _120_   Underground _140_

Water Supply

Water Flow Test	Pump Data	Tank or Reservoir
Date & Time _10/90 10 a.m._	Rated Capacity _____	Capacity _____
Static PSI _107_	At PSI _N.A._	Elevation _N.A._
Residual PSI _98_	Elevation _____	**Well**
GPM Flowing _1950_		
Elevation _0-Ft._		Proof Flow _____ GPM

Location _Hydrant north of new building_

Source of Information _Factory Mutual flow test_

Commodity Storage

Commodity _____   Class _____   Location _____

Storage Height _N.A._   Area _____   Aisle Width _____

Storage Method: Solid Piled _____ %   Palletized _____ %   Rack _____ %

Rack

☐ Single Row   ☐ Conventional Pallet   ☐ Automatic Storage   ☐ Encapsulated

☐ Double Row   ☐ Slave Pallet _N.A._   ☐ Solid Shelving   ☐ Non-

☐ Multiple Row   ☐ Open   Encapsulated

Flue Spacing in Inches	Clearance From Top of Storage to Ceiling
Longitudinal _____ Transverse _____ _N.A._	_____ Ft. _____ In.

Horizontal Barriers Provided _____

A.L.V. —Alarm Valve	G.P.M. —Gallons Per Minute	P.S.I. —Pounds Per Square In.
C —Cross	GV —Gate Valve	PT —Total Pressure in P.S.I.
CV —Swing Check Valve	LT.E. —Long Turn Elbow	Q —Flow Increment
DEL.V. —Deluge Valve	PE —Pressure Loss Due	Q —Summation of Flow
DPV —Dry-Pipe Valve	to Elevation	ST —Strainer
E —90° Elbow	PF —Pressure Loss Due	T —Tee
EE —45° Elbow	to Friction	T.W. —Thin Wall Pipe

051678

Form TS-FC-2
(Page 1 of 3)

Hydraulic Calculation Sheet # 1

Figure 8.6

17/32" orifice   K=8.2   .25 GPM/S.F. over 2800 S.F.   100 S.F./Head

Nozzle Type & Location	Flow In G.P.M.	Pipe Size	Fitting & Devices	Pipe Equiv. Length	Friction Loss P.S.I./Ft.	Required P.S.I.	Hyd. Ref. Pt. ⬡	Elev.	Notes
Line # 1	q			LGTH. 10.00		PT 9.30			Q = K√P
①		1"		FTG.	.199	PF 1.99			√P = Q/K
	Q 25.00			TOT. 10.00		PE			√P = 25/8.2
②	q 27.55			LGTH. 10.00		PT 11.29			P = 9.30
		1¼"		FTG.	.210	PF 2.10			
	Q 52.55			TOT. 10.00		PE			Q = K√P
③	q 30.01			LGTH. 5.0		PT 13.39			Q = 8.2√11.25
		1½"	T	FTG. 8.0	.324	PF 4.21			Q = 27.55
	Q 82.56			TOT. 13.0		PE			Q = K√P
	q			LGTH.		PT 17.60			Q = 8.2√13.3
				FTG.		PF			Q = 30.01
	Q			TOT.		PE			
	q			LGTH.		PT			Q = K√P
				FTG.		PF			K = Q/√P
	Q			TOT.		PE			
	q			LGTH.		PT			K = 82.56/√17.60
				FTG.		PF			
	Q			TOT.		PE			K = 19.66
	q			LGTH.		PT			"K" for
				FTG.		PF			Line #1
	Q			TOT.		PE			19.66
	q			LGTH.		PT			
				FTG.		PF			
	Q			TOT.		PE			
	q			LGTH.		PT			
				FTG.		PF			
	Q			TOT.		PE			
	q			LGTH.		PT			
				FTG.		PF			
	Q			TOT.		PE			
	q			LGTH.		PT			
				FTG.		PF			
	Q			TOT.		PE			
	q			LGTH.		PT			
				FTG.		PF			
	Q			TOT.		PE			
	q			LGTH.		PT			
				FTG.		PF			
	Q			TOT.		PE			
	q			LGTH.		PT			
				FTG.		PF			
	Q			TOT.		PE			
	q			LGTH.		PT			
				FTG.		PF			
	Q			TOT.		PE			

051678

Form TS-FC-2
(Page 2 of 3)

Hydraulic Calculation Sheet # 2

Figure 8.6 (_continued_)

Nozzle Type & Location	Flow In G.P.M.	Pipe Size	Fitting & Devices	Pipe Equiv. Length	Friction Loss P.S.I./Ft.	Required P.S.I.	Hyd. Ref. Pt.	Elev.	Notes
Line #2 ④	q / Q 25.00	1"		LGTH 10.00 / FTG / TOT 10.00	.199	PT 9.30 / PF 1.99 / PE			$Q=K\sqrt{P}$ / $\sqrt{P}=9/K$ / $\sqrt{P}=25/8.2$
⑤	q 27.55 / Q 52.55	1¼"		LGTH 10.00 / FTG / TOT 10.00	.210	PT 11.29 / PF 2.10 / PE			P=9.30 / $Q=K\sqrt{P}$
⑥	q 30.01 / Q 82.56	1½"		LGTH 10.00 / FTG / TOT 10.00	.324	PT 13.39 / PF 3.24 / PE			$Q=8.2\sqrt{11.29}$ / Q=27.55 / $Q=K\sqrt{P}$
⑦	q 33.46 / Q 116.02	2"	T	LGTH 5.00 / FTG 10.00 / TOT 15.00	.125	PT 16.63 / PF 1.88 / PE			$Q=8.2\sqrt{13.3}$ / Q=30.01 / $Q=K\sqrt{P}$
Line #1 R.P.	q 84.54 / Q 200.56	2½"	T	LGTH 2.00 / FTG 12.00 / TOT 14.00	.147	PT 18.51 / PF 2.06 / PE .87			$Q=8.2\sqrt{16.63}$ / Q=33.46 / $Q=K\sqrt{P}$
CM "A"	q / Q 200.56	2½"		LGTH 10.00 / FTG / TOT 10.00	.147	PT 21.44 / PF 1.47 / PE			$Q=19.66\sqrt{18.}$ / Q=84.54
Lines 3+4 CM "B"	q 207.50 / Q 408.06	3"		LGTH 10.00 / FTG / TOT 10.00	.189	PT 22.91 / PF 1.89 / PE			"Q" For Line #1 = 84.54 / PE=(2.0)(.434)
Lines 5+6 CM "C"	q 215.73 / Q 623.79	3½"		LGTH 10.00 / FTG / TOT 10.00	.203	PT 24.80 / PF 2.03 / PE			$Q=K\sqrt{P}$ / $K=\frac{Q}{\sqrt{P}}$
Lines 7+8 CM "D"	q 224.40 / Q 848.19	3½"		LGTH 6.50 / FTG / TOT 6.50	.359	PT 26.83 / PF 2.33 / PE			$K=\frac{200.56}{\sqrt{21.44}}$ / K=43.32
BM "E"	q / Q 848.19	3½"		LGTH 33.50 / FTG / TOT 33.50	.359	PT 29.16 / PF 12.03 / PE			"K" Lines 1+2 / $Q=K\sqrt{P}$ / $Q=43.32\sqrt{22.9}$
BM "F"	q / Q 848.19	4"	2 E	LGTH 111.00 / FTG 20.00 / TOT 131.00	.194	PT 41.19 / PF 25.41 / PE			Q=207.50 / $Q=K\sqrt{P}$ / $Q=43.32\sqrt{24.8}$
Riser	q / Q 848.19	4"	AL.VA.+OS&Y	LGTH 24.50 / FTG 22.00 / TOT 46.50	.194	PT 66.60 / PF 9.02 / PE 10.63			Q=215.73 / $Q=K\sqrt{P}$ / $Q=43.32\sqrt{26.8}$
Inside Hose	q 150.00 / Q 998.19			LGTH / FTG / TOT		PT 86.25 / PF / PE	Base Riser		Q=224.40 / PE=(24.5)(.434)
Undg. F.S. Pc.	q / Q 998.19	6"	E+4×H FL6	LGTH 6.00 / FTG 19.00 / TOT 25.00	.036	PT 86.25 / PF .90 / PE 2.60			PE (6.0)(.434)
Undg. to conn. City main	q / Q 998.19	6"	T+Gate VA	LGTH 100.00 / FTG 36.00 / TOT 136.00	.036	PT 89.75 / PF 4.99 / PE			
Outside Hose	q 500.00 / Q 1498.19			LGTH / FTG / TOT		PT 94.74 / PF / PE			
	q / Q			LGTH / FTG / TOT		PT / PF / PE			

1500 GPM @ 95 PSI

Hydraulic Calculation Sheet #3

Figure 8.6 (continued)

Figure 8.6 (continued)

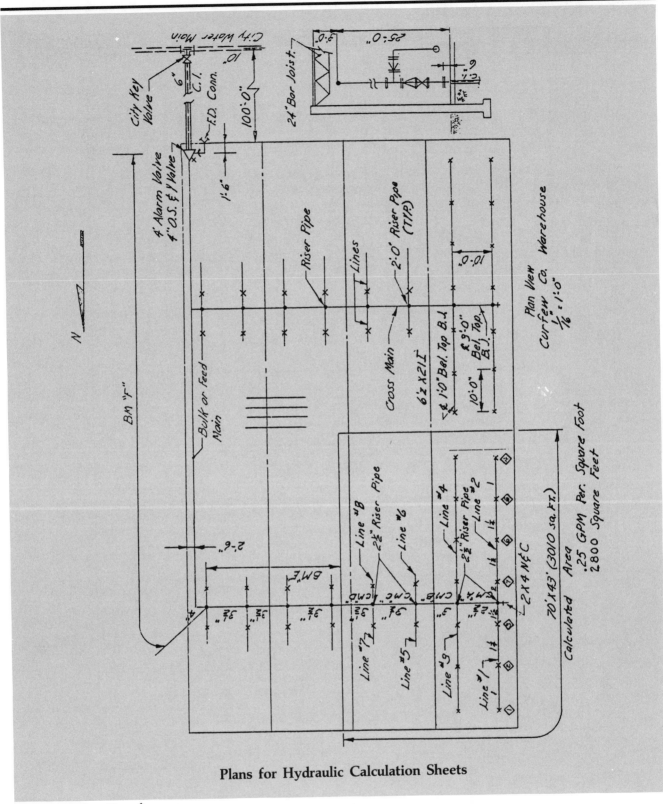

Plans for Hydraulic Calculation Sheets

Figure 8.6 (continued)

Explanation of Hydraulic Calculation Sheets—1 through 4

A. Sheet One: Cover or Information Sheet.

1. Basic Design Data: This criteria is, in most instances, furnished by the insurance interest having jurisdiction. If density and area are obtained from NFPA pamphlet No. 13, or FM Loss Prevention Data, it is imperative that the criteria be approved by the insurance interest having jurisdiction prior to preparing shop drawings.
2. Flow Test Data—See Section VII, Margins of Safety. The supply and demand curves (Sheet 4) present a graphic indication of the importance of a realistic and accurate flow test.
3. The balance of Sheet 1 is self-explanatory and furnishes a concise overview of the entire hydraulic situation.

B. Calculations:

1. Design Criteria: .25 gpm per-square-foot area of application—2,800 square feet.
2. Due to construction of the building, the system has been laid out on the basis of 10 feet between heads, 10 feet between lines: 100 square feet per head.
3. An area of 2,800 square feet is selected most remote from the supply.
   a. If the design criteria can be met for this area, there is assurance that areas closer to the supply (shorter supply runs) will be well within the design criteria.
   b. We have picked the "worst condition"—hydraulically—due to the longest supply to feed this area, and is furthest from the system supply.
4. Step No. 1: (2,800 divided by 100 sq. ft./head = 28 heads calculated)
   a. 100 square feet per head .25 gpm per square foot
      $100 \times .25 = 25$ gpm.
   b. This means that the furthest sprinkler head from the supply must deliver 25 gpm.
   c. Head 1 (nozzle type and location column).
   d. Q = 25 gpm (flow in gpm column).
5. Step No. 2:
   a. Now we must determine what pressure will be required at head 1 to deliver 25 gpm.
      (1) In order to keep the pressure in the total system within the flow test range, and keep pipe size to a minimum, and in view of a good gpm flow, we have chosen a "large orifice" head which gives a large volume at a low pressure.
      (2) From the manufacturer's data, the L.O. head we are using (Viking) has a "K" factor of 8.2.
      (3) The "K" factor is a constant which reduces the friction loss in the head, and fitting the head screws into, to a single number for loss through a single orifice.
   b. Using the Formula: $Q = K\sqrt{P}$
      Q = gpm flow
      K = Constant
      P = psi

Figure 8.7

Explanation of Hydraulic Calculation Sheets—1 through 4

And substituting

$$Q = K \sqrt{P}$$
$$\sqrt{P} = Q/K$$
$$\sqrt{P} = 25/8.2$$
$$P = 9.30 \text{ psi}$$

It requires 9.30 psi at the head to deliver 25 gpm. Insert 9.30 in "required psi" column, PT = Total Pressure.

6. Step No. 3:
   a. We select 1-inch pipe size for the end head supply and enter this in "Pipe Size" column.
      (1) This may have to be changed if the friction loss is too great.
   b. There are no fittings to enter in the "Fitting and Device" column since the "K" factor takes into account the fitting supplying the head.
   c. We enter 10.00 feet in "pipe equiv. length" column since the pipe supplying head 1 is 10 feet long.
   d. This 1-inch pipe must deliver 25 gpm to head 1—the friction loss per foot tables for 1-inch pipe, 25 gpm flowing, is .199 psi/ft. $10 \times .199 = 1.99$ psi (total friction loss)
      Enter this in "required psi" column under "PF" (friction loss psi).
   e. The pipe is run level (NFPA 13, 1978, Wet System Piping Drainage) so there is no static, or elevation, loss to enter after "PE" (elevation psi loss or gain).

7. Step No. 4:
   a. Total the psi required (to deliver 9.30 psi at end head) at head 1 and friction loss: PT = 11.29
   b. At head 2 there is a total of 11.29 psi and we must determine how many gpm 11.29 psi will deliver.

   $$Q = K \sqrt{P}$$
   $$Q = 8.2 \sqrt{11.29}$$
   $$Q = 27.55$$

   c. Head 2 will deliver 27.55 gpm and head 1 is delivering 25.00 gpm.
   d. Enter head 2 delivery in flow column under small "Q" and add to head 1 flow.
   e. Total "Q" is now 52.55 gpm and the second piece of pipe must deliver 52.55 gpm.
   f. Select 1-1/4-inch pipe. Length 10.00 feet. No fitting.
   g. Friction loss per foot—1-1/4-inch pipe flowing 52.55 gpm is .210 psi/ft. $10 \times .210 = 2.10$ psi friction loss—enter in "PF" and add to obtain a new total "PT" of 13.39 psi.

8. Step No. 5:
   a. Determine what gpm head 3 will deliver at 13.39 psi.

   $$Q = K \sqrt{P}$$
   $$Q = 8.2 \sqrt{13.39}$$
   $$Q = 30.01 \text{ gpm}$$

   b. Enter this after small "Q" and add to gpm for head 1 and 2.
   c. Total gpm that the third piece of pipe must deliver—large "Q"—is 82.56 gpm.
   d. Select 1-1/2-inc pipe. Length is 5 feet to the tee on top of the riser pipe. (The riser pipe supplies lines number one and two).

Figure 8.7 (continued)

Explanation of Hydraulic Calculation Sheets—1 through 4

    e. The water coming up the riser pipe must change direction in the tee on top to enter line number one.

    f. To obtain the friction loss due to this change in direction we look on the "Equivalent Pipe Length Chart." Under "tees" for 1 and 1/2-inch (use the pipe size of the pipe entering the tee) the friction loss in the tee is 8 feet. Enter this 8-feet in "ftg" slot and add to pipe length. Total is 13-feet.

    g. From the table 82.56 gpm flowing through 1-1/2-inch pipe—.324 psi friction loss per foot.
        $13 \times .324 = 4.21$ psi (Total Friction Loss).
        Enter this in "required psi" column under "PF".

    h. The total pressure at the centerline of the riser pipe tee is 17.60 psi.

9. Step No. 6:

    a. The same process is performed for line number two (Sheet 3).

    b. We find that at the centerline of the riser pipe tee the following condition exists.
        Line (1) 82.56 gpm @ 17.60 psi.
        Line (2) 116.02 gpm @ 18.51 psi.

    c. Since it is impossible to have two pressures at this point, and since the minimum pressure to supply line number is 18.51 psi, we must determine what gpm line one will receive at 18.51 psi in lieu of 17.60 psi.

    d. To do this we must convert line one to a "K" factor.
$$Q = K \sqrt{P}$$
$$K = Q \sqrt{P}$$
$$K = 82.56 \sqrt{17.60}$$
$$K = 19.66 \text{ (K for line } \# 1)$$

    e. Now we can determine what this "single orifice" (line one) will receive (gpm) at the higher psi.
$$Q = K \sqrt{P}$$
$$Q = 19.66 \sqrt{18.51}$$
$$Q = 84.54 \text{ gpm (Flow to line 1 at 18.51 psi.)}$$

    f. At a balanced pressure (18.51 psi) at the centerline of the tee on top of the riser pipe ($2 \times 1\text{-}1/2 \times 2\text{-}1/2$) the gpm to line (1) is 84.54 gpm and to line (2) is 116.02 gpm.

10. Step No. 7:

    a. The riser pipe ("R.P." Sheet 3) must supply line one and line two.

    b. Enter line on gpm in "flow in gpm" column (Sheet 3) and add to line two flow. Total "Q" = 200.56 gpm.

    c. Select 2-1/2-inch for riser pipe size.
      (1) Once again—this is a matter of judgment, taking into consideration total flow vs. friction loss.

    d. The length of the riser pipe is 2 feet.
      (1) Section shows 3 feet to the cross main and plan indicates centerline of lines 1-foot below top of B.J. Therefore, 2-foot riser pipe.

    e. The water from the cross main must change direction to enter the riser pipe, so we have a fitting friction loss to consider. From the table 2-1/2" tee = 12 feet.
        2 feet + 12 feet = 14-foot total, for riser pipe.

    f. 200.56 gpm flowing through 2-1/2-inch pipe—.147 psi friction loss/foot $14 \times .147 = 2.06$ psi friction loss in riser pipe (PF = 2.06).

Figure 8.7 (*continued*)

     g. There is a 2-foot elevation change—loss, since the water must rise 2 feet from the cross main to lines one and two.

        $2 \times .434 = .87$ psi static loss

        PE = .87

     h. Adding PT at top of riser pipe to PF and PE = a PT of 21.44 psi at the centerline of the cross main tee feeding the riser pipe.

11. Step No. 8:

     a. Cross main "A" must supply 200.56 gpm.

     b. Length cross main (CM) "A" = 10 feet.

     c. Friction loss in selected pipe size for CM "A", 2-1/2-inch, with 200.56 gpm flowing is .147 psi/ft.

        $10 \times .147 = .147$ psi.

     d. 21.44 psi (PT) plus PF 1.47 psi = 22.91 psi (PT at centerline of cross main tee feeding riser pipe and lines three and four.

12. Step No. 9:

     a. Lines three and four supplying riser pipe are all identical to lines one and two and R.P. for one and two.

     b. Necessary to determine the gpm lines three and four will receive at a psi of 22.91.

     c. Line one and two and R.P. must be reduced to a "single orifice" flow ("K").

        $Q = K\sqrt{P}$

        $K = Q\sqrt{P}$

        $K = 200.56\sqrt{21.44}$

        K = 43.32 ("K" at centerline of cross main tee for R.P. to lines one and two).

     d. Substituting

        $Q = K\sqrt{P}$

        $Q = 43.32\sqrt{22.91}$

        Q = 207.50 gpm ("Q"—flow—to lines three and four @ 22.91 psi).

13. Step No. 10:

     a. CM B must supply lines one and two (200.56 gpm) and lines three and four (207.50 gpm), or a total of 408.06 gpm.

     b. Select 3-inch pipe for CM B.

     c. Length CM B = 10 feet.

     d. Friction loss with 408.06 gpm—3-inch pipe = .189 psi/ft.

        $10 \times .189 = 1.89$ psi (PF)

     e. Total (PT) at centerline of CM tee for lines five and six is 24.80 psi.

14. Step No. 11:

     a. Lines five and six and R.P. are identical to lines one and two and R.P. which we reduced to a "K" of 43.32.

     b. At 24.80 psi lines five and six will receive:

        $Q = K\sqrt{P}$

        $Q = 43.32\sqrt{24.80}$

        Q = 215.73 gpm.

15. Step No. 12:

     a. CM "C" must now supply a total of 408.06 gpm plus 215.73 gpm = "Q" of 623.79 gpm.

     b. CM "C" = 10-foot length.

     c. Select 3-1/2-inch pipe size for CM "C".

     d. 623.79 gpm flowing through 3-1/2-inch pipe—friction loss—.203 psi/ft.

        $10 \times .203 = 2.03$ psi (PF)

Figure 8.7 (continued)

    e. Total pressure (PT) at centerline of CM tee supplying line seven and eight is 26.83 psi.

    f. Lines seven and eight and R.P. identical lines one and two and R.P. "K" for lines one and two and R.P.—43.32.

$$Q = K \sqrt{P}$$
$$Q = 43.32 \sqrt{26.83}$$

       Q = 224.40 gpm delivered to lines seven and eight @ 26.83 psi.

    g. CM "D" must supply 623.79 gpm plus 225.50 gpm = total "Q" = 848.19 gpm.

    h. CM "D" length—to the perimeter of the calculated area is 6.5 feet.

    i. Select 3-1/2-inch pipe for CM "D".

    j. 848.19 gpm through 3-1/2-inch pipe—.359 psi/ft friction loss. 6.5 × .359 = 2.33 psi (PF).

    k. Total psi (PT) at perimeter of calculated area is: 29.16 psi with 848.19 gpm flowing.

16. Step No. 13:

    a. It has been determined that we must supply the "most remote" (to the supply) area with 848.19 gpm @ 29.16 psi.

    b. It can be easily seen that if we accomplish this criteria we will more than satisfy the design criteria (.25 gpm/sq ft) as we approach the supply.

    c. Therefore: we can consider the CM and bulk main as supply pipe or bulk main (BM).

    d. No friction loss for riser pipe tees or tee that feeds the second CM since the water flows "straight through" with no friction loss.

    e. From the perimeter of the calculated area to the last line tee on the CM is 33.5 feet—BM "E".

    f. Select 3-1/2-inch pipe.

    g. 848.19 gpm flowing through 3-1/2-inch pipe (to the calculated area)—.359 psi/ft friction loss. 33.5 × .359 = 12.03 psi "PF".

    h. "PT" at perimeter of calculated area, 29.16 plus 12.03 = 41.19 psi "PT."

    i. Since BM "F" is long and friction loss would be considerable, select 4-inch pipe.

    j. BM "F" = 111 feet. ELL at north end and at riser (supply riser) where water will change direction. From table 4" ELL = 10 feet pipe equiv. BM "F" = 111.0' plus 20' = 131 feet.

    k. 848.19 gpm flowing through 4-inch pipe—.194 psi/ft friction loss. 131 × .194 = 25.41 psi "PF."

    l. "PT" 41.19 psi plus 25.41 psi "PF" = 66.60 psi at top of supply riser.

17. Step No. 14:

    a. Riser is 24.5 feet high from the flange of the underground to the ELL at the top.

    b. The Fire Department tee is a straight through flow—no PF loss.

    c. The alarm valve and gate valve (OS&Y), 4-inch size, has an equivalent friction loss of 22 feet of 4-inch pipe. 24.5 plus 22 = 46.5 feet.

    d. 848.19 gpm through 4-inch pipe—.194 psi/ft friction loss.

    e. 46.5 × .194 = 9.02 psi "PF."

Figure 8.7 (continued)

    f. Riser is 24.5 feet high.
       Static Loss = 24.5 × .434 = 10.63 psi "PE."
    g. Total psi (PT) at underground flange is 86.25 psi.

18. Step No. 15:
    a. Insurance interest having jurisdiction has allowed us to add in the flow of one 1-1/2-inch hose operating (fed by 1-inch pipe from the sprinkler system) at the base of the riser in lieu of "picking it up" as we calculated back.
    b. 848.19 gpm ("Q") plus 150 gpm (inside hose "Q") = 998.19 gpm.

19. Step No. 16:
    a. 4-inch underground supply to fire extinguishing systems is no longer permitted by insurance services, Industrial Risk Insurers (IRI) or FM, so we must run 6-inch to the City main.
    b. The "F.S.Pc." (flange and spigot piece) is 6 feet long and has a standard 6 × 11-inch flange above the floor.
    c. To convert to the 4-inch riser we will use a 4 × 11-inch reducing flange to connect.
    d. Equivalent pipe length for the 6-inch ELL on the bottom of the F.S.P.C. plus the 4 × 11 flange = 19 feet of 6-inch pipe.
    e. 6 feet plus 19 feet = 25.0 feet.
    f. 998.16 gpm flowing through 6-inch pipe - .036 psi/ft friction loss.
       25 × .036 = .90 psi "PF."
    g. F.S.P.C. is 6 feet high (long).
       PE = 6 × .434 = 2.60 psi static loss (PE).
    h. PT = 89.75
    i. 100 feet 6-inch to City main.
    j. 6-inch gate valve (City key valve-tapping valve) plus City main tee (10 × 10 × 6) equivalent 6-inch pipe length 36 feet.
    k. 998.16 gpm flowing through 6-inch pipe—.036 psi/ft friction loss.
       .136 × .036 = 4.99 psi PF
    l. 89.75 psi "PT" plus 4.99 psi PF = 94.74 psi "PT."

20. Step No. 17:
    a. Add 500 gpm for use of a hydrant while the sprinkler system is in use 998.16 + 500 = 1498.19 gpm.
       Summary: Demand 1500 gpm @ 95 psi.

Figure 8.7 (continued)

CHAPTER NINE

DETECTION AND ALARM SYSTEMS

CHAPTER NINE

DETECTION AND ALARM SYSTEMS

Selecting the proper fire detection and alarm systems requires a careful evaluation of all that the system is designed to accomplish, as well as the hazard and the environment in which the detection system must operate. Detection system selection must fulfill two basic requirements. It must be sensitive enough to detect combustion within an acceptable and required time frame, and it must be reliable enough to not respond to non-fire activation sources. A determination of the proper detection system to use depends on a basic knowledge of the function of the different detectors, the hazard characteristics, and the four stages of combustion. These four stages of combustion are:

1. The **incipient stage**, where invisible products of combustion are produced. At this stage, smoke is not visible, nor has the heat of combustion developed.
2. The **smoldering stage** occurs when the combustion has developed to the point where combustion products are visible as smoke, but flame and heat are not yet a factor.
3. The **flame stage** is an intermediate stage where considerable heat is not yet being produced by the combustion, although this stage very quickly moves into the next, the heat stage. In this stage there may be a clean flame, but little or no smoke.
4. The **heat stage** combines all the elements of combustion— invisible products as well as visible smoke, flame, and heat.

The potential hazards should be thoroughly evaluated in order to determine the anticipated combustion stages and how quickly the combustion will pass through the various stages. This information is integral to the selection of the proper and most efficient detector for the hazard area.

Thermal Detection Systems

The operation of thermal detectors is the least sophisticated of all the detectors covered in this chapter. Thermal detectors are designed to actuate at a predetermined fixed temperature, and they should, therefore, be used when the protected hazard area has the potential to produce considerable combustion heat, or at least sufficient combustion heat to activate the detector before considerable damage is done passing through the early stages of the combustion.

A sprinkler head is actually a thermal detector. When combustion heat reaches the temperature setting of the fusible element of the sprinkler head, the fusible element melts, or softens sufficiently to release the head linkage, thereby releasing the discharge. The water movement in the piping actuates a device that transmits an alarm.

The sprinkler head is a good example of a non-restorable detector. Once the head is activated, it must be replaced. Some thermal detectors are non-restorable. A thermal element melts at a predetermined temperature, releasing a mechanical arrangement such as a spring or a plunger which makes or breaks an electric circuit, activating the alarm system. (See Figure 9.1.)

Other thermal detectors are available that activate at a predetermined fixed temperature, but are self-restoring. After the detector has been activated and the combustion heat removed, the detector returns to room temperature and automatically resets itself.

One type of self-restoring thermal detector operates on the principle of a disk made of two different metals, with a curved shape. When the detector is subjected to combustion heat that reaches the fixed operating temperature of the detector, the disk curvature is reversed, transforming from concave to convex shape. This reversal of the disk allows a spring tension plunger to drop, which makes or breaks an electric circuit, thereby activating the alarm or suppression system.

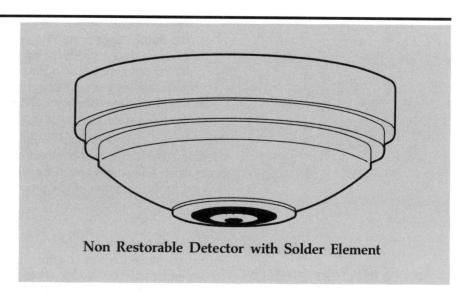

Non Restorable Detector with Solder Element

Figure 9.1

One caution should be noted regarding the use of self-restoring detectors. Unless equipped with special devices, some suppression system valves that are operated when a thermal detector is activated can close when the thermal detector cools and is restored to a normal operating condition. When a suppression system has been activated, it is important that the system not be shut down until the combustion is completely extinguished, and in some instances until the surrounding area has cooled sufficiently to prevent a re-ignition of the combustion. Once the flaming stage of the combustion has been controlled, the temperature at the ceiling could drop below the operating point of the detector and the detector could reset, re-opening or closing a solenoid valve and shutting down the suppression system valve automatically.

When designing a detection system that activates a suppression system, and the detectors are self-restoring, always determine what may be required to maintain the suppression system in an operating condition should the detectors reset. Unless a suppression system is designed to shut down automatically, the system should not be shut down until a professional firefighter determines that it is time to close a manual valve.

Pilot Head Systems

The pilot head detection system is a thermal detection system frequently used to activate suppression systems. The pilot head system uses regular sprinkler heads as the thermal detectors. These heads are installed on 1/2" pipe which is filled with compressed air. When a head fuses, the compressed air is released, and the release of compressed air pressure activates the suppression system valve.

The pilot head system is very reliable in that it is not subject to false trips from non-fire sources, which can cause havoc with other detectors that operate on a principle other than heat.

Pilot head detection systems are frequently used to activate cooling tower and outdoor transformer deluge suppression systems. Since these systems use regular sprinkler heads, the heads must be located so that sufficient heat can collect to fuse the head. When the system is located within a structure, the location of the heads should closely follow the rules for sprinkler head location below a ceiling. If the system is used outdoors, as it would be with transformer protection, each pilot head should have a heat collector installed above the head. This can be a 12" × 12" metal pan securely fastened over each head.

Continuous Line Detector

The continuous line detection system, sometimes referred to as a *continuous wire detection system* or *thermistor system*, uses a detector with the appearance of a wire. A simple description of one type of line detector is a metal tube containing a wire, or conductor, surrounded by a special compound. When a section of the line detector is subjected to the heat of combustion, the compound loses its electrical resistance and a small current passes from the wire in the core of the tube to the metal tube. The control panel senses this current flow and activates the detection system devices. (See Figure 9.2.)

The line-type detector is self-restoring if the fire does not physically damage the equipment. Line detectors come in various lengths joined together at junction points. The line detector can indicate on an annunciator panel the exact location of the line where the combustion heat has activated the line. The line detector is also available in different temperature ranges.

The line detector has many advantages when used in long conveyor enclosures and in conveyors without enclosures, as well as in and around irregularly shaped equipment. It is also used on the ceiling of a building, similar to the placement of spot-type thermal detectors.

The line-type detector has excellent reliability in terms of not responding to non-fire sources of activation, but like any thermal detector, it is not sensitive enough to be referred to as an *early warning detector*, since it must have sufficient combustion heat to react.

Rate of Rise Detector

The rate of rise detector responds to a rate of temperature rise of 15° per minute regardless of the room's ambient temperature. This feature gives the rate of rise detector an advantage over thermal detectors because it is not affected by thermal lag.

When a thermal detector with a fixed operating temperature reaches that operating temperature, the air surrounding the detector is considerably higher than the detector-set temperature. This difference in surrounding air temperature and the operational temperature of the unit is due to the rate at which the temperature rises, and heat is lost in both the body of the unit itself and the surrounding area.

The rate of rise detector is activated by the very rate of temperature rise that is lost with a thermal detector. If the temperature rises at the rate of 15° per minute, or faster than 15° per minute, the rate of rise detector will activate.

Certain disadvantages of the rate of rise detector must be considered. Slow-burning or smoldering combustion may produce considerable heat, and the ceiling where the detector is located may rise to a very high temperature, but unless the rate of heat rise is rapid enough to produce a 15° per minute increase, the detector will not activate. It is for this reason

Stainless Steel Outer Jacket

Conductor

Ceramic Thermistor Core

Continuous Wire Detector (Thermistor)

Figure 9.2

228

that combination fixed temperature and rate of rise detectors are used to overcome the rate of temperature rise disadvantage. The combination fixed temperature and rate of rise detector are activated when the rate of temperature rise is 15° or more per minute. It also has a fusible element with a predetermined operating temperature.

The combination electrical spot detector is a hollow shell containing a diaphragm. (See Figure 9.3.) When the shell is subjected to combustion heat that is increasing in temperature at a rate of 15° per minute or greater, the air trapped in the detector shell expands, thereby forcing the diaphragm up to make an electrical contact that closes a current and activates the system.

The system may be an alarm system, a suppression system, an equipment shut-down system, an equipment start-up system, or a combination of any of these systems. By closing the electrical circuit, the detector signals the control panel which transmits a signal, or signals, to the various systems.

There is a small vent on the detector to vent small pressure changes within the shell due to changes in the environment where the detector is located. These environmental changes may include starting up the area heating system, or sun on the metal roof after a cool night, or any number of situations that cause a slow rate of temperature increase. These increases are not in the 15° per minute range, but would cause a slow increase in pressure within the detector shell which could eventually depress the diaphragm and activate the detector if not vented.

The vent is calibrated so as not to be capable of venting sufficient pressure increase when the temperature rise of 15° per minute increases the pressure to the trip point.

Rate-of-Rise and Fixed Temperature Detector

Figure 9.3

Except for the fusible element that melts and is released from the detector at a fixed temperature, the detector in the rate of rise mode is self-restoring. When the fusible element reaches its operating temperature and is released from the detector, it releases a spring that mechanically depresses the diaphragm and activates the detector. (See Figure 9.3.)

Heat Actuating Device

The heat actuating device, commonly referred to as the H.A.D., is a rate of rise detector that has been on the market and in service for many years. Basically, the H.A.D. is a hollow cylinder connected to the system operating devices through small copper tubing. When there is an increase in combustion temperature, the air in the H.A.D. expands and the pressure created is transmitted through the copper tubing to activate a diaphragm, or equivalent device, that reacts to this increased pressure.

This detection system is pneumatic, and as a result, has been used frequently in areas where flammable liquid vapors may be present and an electric spark from an electric detector activation could cause an explosion.

The H.A.D. pneumatic system copper tubing is connected to a device called a *mercury check*, which acts as the environmental temperature change vent for a group of H.A.D.'s that are connected to the mercury check. The mercury check also prevents an increase in pressure in one H.A.D. in the group from being dissipated by pressurizing the other H.A.D.'s in the group. In addition to its environmental temperature change venting function, and its prevention of pressure loss by transfer of pressure from one H.A.D. to another, the mercury check is a retard device to prevent false alarms.

A pneumatic H.A.D. system can be supervised to detect any air leaks in the copper tubing or the H.A.D.'s. A small amount of air pressure can be maintained in the system, and should an H.A.D. develop a leak, or should there be a break in the copper tubing or a leak in the tubing fittings, the loss of pressure will operate a pressure switch or release a diaphragm and activate a trouble signal.

Thermostatic Release

There is a detector on the market that operates on the rate of rise principle when the detector is subjected to a temperature rate of increase of 15° or more per minute. This detector can also be equipped with a sprinkler head to serve as a fixed temperature detector. The detector can thereby function as a pneumatic or hydraulic system.

The thermostatic release detector operates on the principle of two metals expanding at different rates when subjected to temperature change. The long metal tube will expand more rapidly than a rod contained within the tube. The tube expansion exerts a pulling force on the rod, which in turn mechanically operates a valve that releases pressure on a diaphragm, which then raises a clapper and vents the system.

The valve is actually a Schrader valve, the same valve found on the tires of a car or bicycle. When the rod is pulled by the expanding tube, it depresses the stem of the Schrader valve and releases air trapped below the diaphragm. The diaphragm then opens the vent, releasing air when the system is pneumatic, and water when the system is hydraulic.

Ionization Detector

In evaluating the hazard, the level of required detection sensitivity, the environment in which the detector will function, and the type of combustion anticipated, it may be determined that the combustion should be detected in the incipient stage before there is appreciable heat and smoke in order to prevent heat and smoke damage to delicate electronic equipment. In such cases, ionization detectors should be seriously considered because they react when invisible products of combustion are given off by the combustion before there is visible flame, or heat development at the ceiling level, and little or no visible smoke has developed.

The analysis of the combustibles present and the type of fire that these combustibles will produce should be the first concern when making a detector selection. Fires that develop rapidly into the flaming stage without the accumulation of heavy, visible smoke particles are detected most effectively by ionization detectors.

Photoelectric Smoke Detector

A photoelectric smoke detector is activated by visible smoke particles entering the detector. When a smoke detection system is proposed, it is recommended that a determination be made as to whether ionization detectors or true smoke detectors would be best suited to the hazard conditions. It is not just a matter of terminology, but is important to differentiate between ionization and the photoelectric detector when recommending either type. The general public generally refers to both the ionization and the photoelectric detector as "smoke detectors," but when an engineer is recommending a detection system based on the results of an engineering evaluation, it is very important that the terminology be correct.

The photoelectric detector responds to the thicker, heavier, visible smoke that develops when slow-developing and smoldering combustion is present. A slow-burning or smoldering fire is typical of combustion of materials found in the office, home, hotels, dormitories, and hospitals, since this type of combustion produces the type of smoke created by this type of occupancy.

It is important to realize that the selection of the proper detector is an extremely difficult decision because of the many variables that sometimes make it almost impossible to determine with accuracy the nature of the potential fire. Every modern building contains equipment and furnishings manufactured of different, and sometimes unique, materials that may produce a type of combustion that is not anticipated in evaluating the proper detector for the area. Although the cost of a detection system will be increased, the engineer may, in order to provide reliable combustion detection, recommend the installation of both ionization and photoelectric detectors if the type of fire potential cannot be readily determined.

In an effort to obtain a sensitive, early warning detection system where ionization detectors appear to be the best selection, it is imperative that the general environment be considered because these detectors will react to many non-fire invisible particles, including exhaust fumes, welding operations, kitchen odors, and pilot lights. Almost any detector can be activated by electrical equipment of a certain voltage and radiation level. Hospitals may have problems with detectors exposed to CAT scanners, high frequency radio transmissions, etc.

Maintenance is a prime consideration with detection systems. Dust and contamination can affect the sensitivity of any detector, either by making it more sensitive to non-fire detection sources, or by reducing the sensitivity.

A brief description of the way in which the ionization and the photoelectric detectors function should help in selecting the proper detector for a given situation.

The ionization detector contains a minute amount of radioactive material that ionizes the detector chamber. In other words, the radioactive material emits sufficient material to charge the atmosphere in the chamber of the detector to develop a current flow through the air space separating two electrodes. When minute invisible products of combustion (and/or small smoke particles) enter this electrically charged chamber, they attach themselves to the charged radioactive particles in the current flow, and by so doing, reduce the current flow between the two plates.

When the flow level between the plates is reduced by these "heavy" particles to a predetermined level, the minute voltage change is amplified and the detector is activated. This is a simplified explanation of how the ionization detector functions, and is offered to further illustrate the difference between an ionization detector and a photoelectric smoke detector. A photoelectric smoke detector works on an entirely different principle than the ionization detector. One type of photoelectric smoke detector uses a light source that is projected onto a photosensitive device. When visible smoke particles enter the detector and pass between the light source and the photo-sensitive receiver device, they reduce the light reaching the receiver. This causes a slight voltage change, which is then amplified, and the detector is activated.

Another type of photoelectric detector uses the light source and the photosensitive light receiver, but with this particular detector, the light source does not fall on the receiving sensor. When visible smoke particles enter the detector, the light is scattered by the smoke particles and in addition, the light is reflected off the smoke particles. When enough of this light reflected strikes the sensor receiver, the voltage change is amplified and the detector activates.

Linear Beam Smoke Detector

The linear beam smoke detector basically operates on the same principle as the photoelectric smoke detector with a light source directed on a photo-sensitive receiver. The difference is that the linear beam smoke detector consists of two separate units. The light source is one piece of equipment, and the light beam receiver is a separate piece of equipment. The linear beam smoke detector light source, or transmitter, emits an invisible infrared beam over open area distances of 35' to 300' onto the light receiver. When no smoke interferes with this beam, the receiver accepts the beam at a specified voltage level, but when smoke interferes with the beam, the infrared light reaching the receiver is lessened, and when the beam intensity drops below the predetermined sensitivity level of the receiver, it initiates a signal from the detector. (See Figure 9.4.)

This type of smoke detector has many applications in areas where spot-type smoke detectors could not be installed to operate efficiently, such as atriums, airport terminals with very high ceilings, aircraft hangars, churches, and large open facilities. Spot-type smoke detectors installed on the extremely high ceilings of these areas would be almost inaccessible for maintenance, and since the beam transmitter and the sensor receiver are installed on the walls, the architectural beauty of a church ceiling or hotel atrium will not be compromised by the appearance of several spot detectors. One special application of a linear beam detector is in psychiatric facilities and correctional institutions where spot-type detectors would be subject to vandalism.

Rate Compensation Detector

The rate compensation detector, sometimes referred to as a *rate anticipation detector*, consists of two metallic struts mounted inside of a steel elongated shell.

The stainless steel shell has a coefficient of expansion greater than the two metallic struts mounted within it. When the shell is exposed to heat, it expands, and since the ends of the two struts are attached to either end of the shell, the expanding shell pulls the struts and stretches them until they make contact at the center of the struts.

Rate compensation detectors have a predetermined fixed temperature set point, and when the two struts make contact and activate the detector, it has reached this temperature set point. Unlike the rate of rise detector that requires a minimum of 15° per minute rate of temperature rise to operate,

Linear Beam Smoke Detector

Figure 9.4

a smoldering fire producing a slow rise in temperature will heat both the outer shell and the struts equally until the set point of the detector is reached.

When the combustion heat rate is rapid, the rate compensation shell expands rapidly until the struts are stretched to the point of contact, thus eliminating the thermal lag that may prevent a fixed temperature detector from operating when the surrounding air has reached the set point of the detector.

Duct Detector   Duct detectors are either ionization or photoelectric. They are usually installed on a return air duct close to the fan unit. Duct detectors are installed on the exterior of the duct with long tubes extending into the duct. These tubes have holes that pick up products of combustion in the air stream returning from the vents in the facility, transmitting them into the detector unit on the outside of the duct. (See Figure 9.5.)

When the detector is activated, one of the signals it actuates is an automatic shut-down of the heating and air-conditioning fans to prevent the products of combustion from being sent back through the heating and air-conditioning system throughout the building.

Duct detectors provide an excellent function, but always keep in mind that they should not substitute for a detection system throughout the facility. The return air to the fan units may come from many parts of the building, and smoke generated in one room of the building and sucked into the return air vent of that room may be so diluted by return air from other rooms where no combustion is occurring that by the time it reaches the duct detector, it is not concentrated enough to activate the detector.

Ionization
Smoke Detector

Duct–Mounted
Smoke Detector

Photoelectric
Smoke Detector

Smoke Detectors

Figure 9.5

Flame Detector

Flame detectors respond to the visual radiant energy of a flame. A flame produces visible wavelengths of radiant energy, and different flame detectors see different radiant wavelengths given off by a flame. There are two basic types of flame detectors. One is ultraviolet, and the other is infrared. The ultraviolet flame detector responds to ultraviolet flame wavelengths, and the infrared responds to infrared flame wavelengths.

Ultraviolet flame wavelengths are the shorter wavelengths emitted by high intensity flames, but this range of wavelengths is also emitted by the sun, so when using an ultraviolet detector it must be designed to be "solar blind."

The ultraviolet detector sensitivity decreases with the distance from the flame source, whereas the infrared detector will function effectively when separated some distance from the flame source. This is a generic statement, backed up by the actual fact that ultraviolet detectors installed in an aircraft hangar have been activated by a welding operation several thousand feet from the hangar.

Flame detectors are quite specialized detectors and a considerable level of expertise is required to design a system that is free from activation by non-fire sources, yet sensitive enough to activate at the first flicker of a flame. These detectors are also extremely sensitive, and a system using them must be designed with great care by experienced engineers. A rotating beam light on top of a truck, a slowly rotating fan blade, or the heat from an aircraft engine can activate a flame detector.

In order to attempt to reduce false activation, the ultraviolet and infrared detector are now being used in tandem. Both must be activated in order to activate alarms or trip systems. It is anticipated that a non-fire ultraviolet source will not occur at the same time as a non-fire infrared source, but that a flame in a flammable liquid will produce a flame front that will emit both ultraviolet and infrared wavelengths sufficient to activate the detector. (See Figure 9.6.)

The ultraviolet detector has a viewing area resembling a cone. This cone of vision is approximately 90° closer to the detector, and widens out to approximately 60° at some distance from the detector. The detector sensitivity decreases as the distance from the flame source to the detector increases. For example: A one-square-foot hydrocarbon pan fire can be detected when located some 50' from the detector, and a nine-square-foot hydrocarbon fire can be detected some 100' from the detector.

As the fire size increases, the detector can be activated with the flame front at greater distances from the detector. It must, however, be kept in mind that the sensitivity of the detector is decreased as the flame front moves away from the center line of the detector. This is particularly important when designing floor coverage detection with the ultraviolet detector. The cones of vision of the individual detectors should overlap so that a flame anywhere in the area will be within the sensitive area of the detector cone of vision. (See Figure 9.7.)

The infrared detector also has a cone of vision similar to, but not as wide as, the ultraviolet detector. (See Figure 9.8.) As with the UV detector, the IR detector is most sensitive to flame detection when the flame source is directly in line with the detector. The IR detector has only has an 80% sensitivity when the flame source is 30° off the center of the detector, and 50% sensitivity is accomplished when the flame source is 45° off the center line of the detector.

When flame detectors (optical detectors) are selected for the detection system, and the system is to activate suppression systems, it is recommended that one detector activation operate the alarm system, and the second detector operate the suppression system.

Flame, or optical, detectors are extremely sensitive to flaming fires, but are also very sensitive to non-fire sources of activation. When a flame detection system not only activates an alarm system, but also automatically operates a suppression system, it may be advisable to monitor the optical system for a short period of time after installation.

This monitoring is best explained using the example of aircraft hangar protection: When the foam systems in a large aircraft hangar are activated, they will discharge a large amount of foam over a wide area. This expended foam is expensive to replace, and the labor involved in replacement can also be time consuming and costly. The cost of foam replacement is actually minor compared to the cost impact of discharged foam removal from machinery, electrical equipment, HVAC units, and the aircraft parked in the hangar.

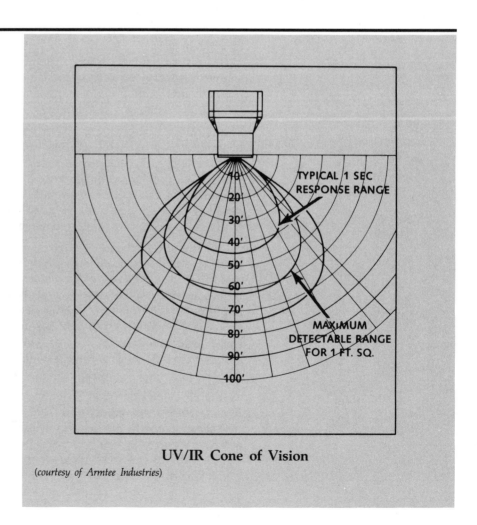

UV/IR Cone of Vision

(courtesy of Armtee Industries)

Figure 9.6

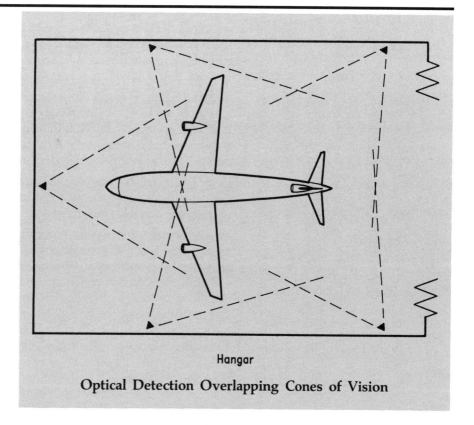

Hangar

Optical Detection Overlapping Cones of Vision

Figure 9.7

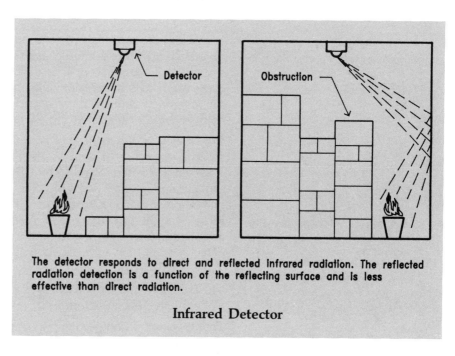

The detector responds to direct and reflected infrared radiation. The reflected radiation detection is a function of the reflecting surface and is less effective than direct radiation.

Infrared Detector

Figure 9.8

The sensitivity of the flame detectors is critical to the protection of the aircraft undergoing maintenance in the hangar, and to the hangar structure itself. A jet fuel spill fire under the aircraft must be controlled by the suppression system within 30 seconds and extinguished within 60 seconds. Otherwise, the aircraft could be damaged beyond repair.

Only an ultraviolet or an infrared detector can "look" under the wings of an aircraft with the sensitivity to trip the suppression systems within the required time frame. But it is this very sensitivity to non-fire sources of activation that can cause the serious and expensive problem of "false dumps" of the suppression systems.

Once the installation of the detection and suppression systems is complete, the detection system is connected to a monitor, having the signals from the detection system interlocked with the suppression systems.

When the detection system is activated, the alarm system becomes operational, and if a fire is observed, the suppression systems can be activated by operation of a manual pull station. The detection system also activates the monitor, which prints out a date and time of the activation. The non-fire source of activation can usually be determined. Having a record of the causes of false detector activations will allow corrective measures to be instituted before the suppression systems are placed on automatic activation. Monitoring has an initial cost, and the manual activation of the suppression systems in a fire situation requires prompt attention. Nevertheless, if the monitoring keeps an area as free as possible from non-fire-related sources of detector activation, the initial cost of the monitoring is negligible compared to that of unnecessary system operation.

Design Criteria

In this section, some design considerations of flame detection systems have been briefly presented. It is extremely important that detection systems that alarm and operate suppression systems be designed with every effort to prevent false activation of the alarm and/or the suppression systems, without decreasing detection sensitivity below the point at which it reacts effectively in a fire situation.

Cross zoning is one method of accomplishing this design criteria. A detection system that is cross-zoned consists of half the detectors classified and wired into the control panel as Zone A, and the other half of the detectors classified and wired into the control panel as Zone B. If one detector in Zone A or Zone B is activated, the alarm system is actuated. If the first detector to operate is in Zone A, then only a detector activated in Zone B will actuate the suppression system. If the first detector to be activated is in Zone B, then only a detector activated in Zone A will actuate the suppression system.

The detectors in a cross-zoned system are spaced so that Zone A and Zone B detectors are intermixed.

Priority matrix is another method of accomplishing the design of a detection system with non-fire activation reliability without losing the required sensitivity. Priority matrix detection systems place all the detectors in one area, or basically, one zone. Any one detector to be activated will activate the alarm system and also form a matrix of detectors surrounding the activated detector. Matrix systems can be related to cross-zoning systems by imagining that when one detector is activated, all the surrounding detectors in the matrix become the same zone classification as the activated detector. Imagine that the activated detector is "Zone A," and the matrix detectors all become "Zone A." (See Figure 9.9.)

Consider the situation where one detector is activated and forms a matrix of surrounding detectors, but a shift in the products of combustion activates a second detector beyond the perimeter of the matrix. This newly activated detector outside of the matrix will not actuate the suppression system, but will form its own matrix of surrounding detectors. Should a second detector in either matrix of surrounding detectors be activated, the suppression system will be activated.

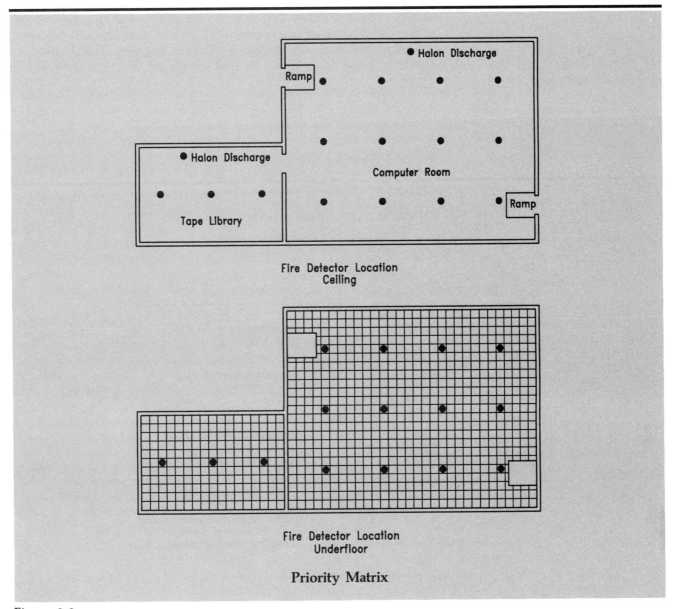

Figure 9.9

There is a detector and detection system on the market that uses *alarm verification* to eliminate nuisance or false alarms. A brief, nontechnical, description of this system is the ability of the detector and control panel to verify a detection source before the system activates an alarm. The detector receives a detection activation source, but it does not go into an alarm condition. Instead, it clears the signal and "waits" to see if the source of activation again activates the detector. If the detector is again activated, the alarm system is actuated. (See Figure 9.10.)

One other detection system to be considered is a system in which the wiring and control panels are arranged so that the first detector activated operates the alarm system, and any second detector activated will actuate the suppression system. This system can be used when there is concern that draft conditions, or some other environmental condition, may not operate a second zone detector in a cross-zoned system, and the same may be true for a second detector in a matrix system, within the time frame required to operate the suppression system.

The selection of the proper detector, and the selection of the most effective type of detection system is a complex engineering judgment call, but there is yet another engineering decision that must be made before specifying, estimating, and designing a detection system. This decision is whether to use a two-wire or four-wire installation, and both of these

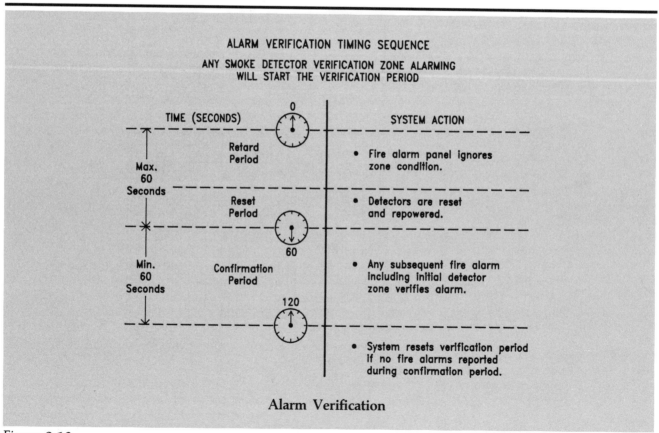

Alarm Verification

Figure 9.10

systems have advantages and disadvantages. If a two-wire system is used, the detectors carry power and the alarm-initiating signal over a single pair of wires, but with a four-wire system, the power is carried over one pair of wires, and the signal is carried over a separate pair of wires. All of the detectors, control panels, annunciators, etc., associated with a two-wire system must be compatible, and must be specifically UL-listed to be compatible

If the system is ever extended, additional equipment or detectors added, or changes to the system are required or desired, the identical equipment, devices, and detectors must be used. Making the system proprietary can increase the cost of any changes because it limits the manufacturers who can bid on the equipment. If the original manufacturer has discontinued making or stocking the particular equipment, the cost impact of a new system divorced from the original system can be high.

The four-wire system does offer much more flexibility when it becomes necessary to make changes or add onto the original system, but the initial cost of the installation is obviously higher because the four-wire system doubles the cost of wiring.

The intention is not to discourage the use of the two-wire, less expensive, system. Evaluation and judgment on the part of the specifier, engineer, and estimator are required to determine where and how the detection system is to be used. If it is obviously a "one-time" detection system installation, budget constraints are imposed, and the client or owner is made aware of and can accept the constraints of the two-wire system, serious consideration should be given to proceeding with this type of system.

Regardless of whether the choice is a two-wire or a four-wire system, it is important to obtain detectors and equipment that are listed for two-wire or four-wire systems.

Supervision

Supervision of detection and alarm systems is a key word in the design of all systems. The word *control* used in the following description of system supervision refers to all of the action that a signal from a detection system may control or activate: suppression system activation, equipment shut-down, equipment start-up, magnetic door closers, etc.

Detection, alarm, and control systems that are electrically operated must have wiring supervised to activate a trouble alarm should a wire break, short out, or become unintentionally grounded, or should a detector malfunction. Usually, the trouble signal activates a light and an audible alarm at the control panel or annunciator, and a silence switch shuts off the audible trouble alarm, but the light will remain on until the trouble is corrected. Most equipment reactivates the audible alarm when the trouble is corrected, and the audible alarm will not be silenced until the silence switch is returned to active position, indicating that the panel or annunciator has been returned to operational condition. Supervision for detection, alarm, and control systems is imperative, because a system that has been rendered inoperable without any indication of its condition, will provide a very serious false sense of protection. Pneumatic detection, alarm, and control systems must have a low air alarm device to activate a trouble alarm when the air pressure drops below operational pressure.

Another system installation feature that must be considered in evaluating the area where the system is to be installed is the possibility of flammable vapors or dust conditions, both of which could produce an explosion when the atmosphere is exposed to an electrical spark. When this condition is encountered, the detectors, wiring, and equipment must be specified as *explosion-proof*.

Explosion-proof equipment and wiring is designed to prevent any spark-producing connections from being exposed to the atmosphere where it could initiate an explosion. Explosion-proof equipment and wiring are more expensive to purchase and install; therefore, it is important to always make this determination when specifying and estimating a detection, alarm, and control system.

Another feature of detection, alarm, and control systems that is extremely important to include in the design, specification, and estimate sheet, is battery back-up. If the system does not have a separate, reliable, and uninterruptible source of power to use when the primary, commercial source of power is lost, the system must have a battery back-up power source. Battery back-up to operate the system and all of its components during loss of primary power must be arranged and equipped to come on line automatically, and within a matter of seconds, when the primary source of power is lost. The batteries should be kept fully charged by an automatic battery charger, and the battery power should be sufficient to operate the entire system for an extended time (which is usually 24 to 48 hours). The control panel or annunciator must have an indicator light which shows that the system is operating on battery back-up, and should also have an indication that the batteries are in a charged condition. On the subject of power supplies, supervision, and wiring of detection, alarm, and control systems, the National Electric Code (NFPA Standard 70) requirements must be followed.

Types of Alarm Systems

Local Alarm System
The local alarm system produces alarms only for the facility it serves, or the premises of the area served. This type of alarm system may be used in schools, hospitals, institutions, and industrial facilities where the alarm is audible, and sometimes also visible, within the building or facility it protects. Notification of the fire department is a human operation by telephone or other human means.

Auxiliary Alarm System
The auxiliary alarm system activates local alarms, but this system is also connected to a municipal fire alarm box so that when the alarm system is activated, the municipal fire alarm box is also activated, which in turn transmits a coded signal directly to fire alarm headquarters. This coded signal received by the fire alarm headquarters identifies the individual fire alarm box transmitting the signal. The fire department can be directed to the specific location of the fire alarm box.

Remote Alarm
This is a very effective fire alarm system which, when activated, automatically transmits a signal to a constantly attended location. "A constantly attended location" is most often the most important facet of any fire alarm system, as is the automatic transmission of this remote alarm signal. Any time the human element is introduced, there can be delays in summoning the fire department which can spell the difference between successful lifesaving operations and extinguishment of the combustion, or loss of life and total destruction of the property.

One relatively simple and inexpensive method of transmitting a remote signal to a remote alarm device is over a leased and dedicated telephone line. The remote alarm device can be located in a fire station, or any facility where someone is in constant attendance.

The control panel should be specified as having contacts for the automatic transmission of remote alarms, and the transmission of trouble alarms, which can also be incorporated into the panel. Usually when the remote signal is transmitted to a municipal fire alarm headquarters or fire station, the authority having jurisdiction must give permission to transmit trouble signals. Usually a municipal fire alarm remote device will receive only alarm signals, since the receipt of the signal and correction of trouble on the system should be handled by the owner of the system. If life safety is a major factor, the municipal fire alarm department may allow trouble signals to be automatically transmitted as well.

If the remote signal is transmitted to a municipal station, a requirement is imposed to have the facility owner sign a contract to have the system's remote alarm tested on a monthly or periodic schedule. This requirement is intended to prevent false alarms, with the resulting responses from the fire department. This additional ongoing cost should be realized by the client or owner of the detection and alarm system.

Proprietary Alarm System

The proprietary alarm system serves a facility or complex of facilities with its own alarm-receiving location. An example is a complex of industrial facilities equipped with detection and suppression systems, where every system activates local alarms and also transmits an alarm to a central, constantly attended area that is a part of, and owned by, the complex. This system is an independent alarm and alarm notification system under one ownership. The alarm-receiving station also receives trouble signals, fire pump signals, etc. Usually the alarm and trouble signals are received by an audible and visual indicator. The signal location is typically indicated on a graphic annunciator, or an annunciator that indicates which system or equipment has transmitted the signal. The response can then be immediate to the area where the signal originated.

Central Station

A central station system is a commercial system where alarm signals are automatically transmitted to a central station that employs experienced personnel whose only occupation is the monitoring of the panels that receive the signals. A central station usually has many clients, and when an alarm or trouble signal is received, the panel indication will inform the operator exactly which client facility is transmitting the automatic signal.

Central station firms receive security and other monitoring alarms, and the operators are trained to interpret the signals, dispatch their personnel to investigate the alarm, and also to notify the fire department and/or police department. Naturally, there is a monthly charge for this central station service, and sometimes these central station companies require that their own equipment be installed to transmit the alarms in order to provide adequate integrity of the equipment and to avoid the possibility of inadequate equipment transmitting numerous false alarms.

CHAPTER TEN

REGULATORY AGENCIES AND AUTHORITIES

CHAPTER TEN

REGULATORY AGENCIES AND AUTHORITIES

Understanding regulatory agencies and the authorities having jurisdiction is key to designing, engineering, specifying, estimating, and installing fire protection systems.

A conversational, if not working, knowledge of fire protection systems and the equipment that comprises those systems is important to all construction and facilities management disciplines. However, this chapter covers a subject that separates the professional and experienced fire protection engineer from the other disciplines involved in fire protection projects.

The guides provided by NFPA, Factory Mutual, and other organizations furnish the specifics of system design, but almost every one of these guides makes the statement in so many words, "subject to the approval of the authority having jurisdiction." Who are the "authorities having jurisdiction?" On some projects they are clearly identified, but on many others this phrase used in a specification indicates the specification writer's uncertainty, in which case the project could be headed for trouble.

The authority having jurisdiction may be the insurance carrier, or the specifications may require that the system be installed "in accordance with state and local codes." These codes may reference ordinances and agencies that can seriously impact a project. Designers should be forewarned that even a very tight specification and reference to compliance with NFPA and/or FM requirements can be destroyed by making blanket overall statements like those mentioned above.

Today's economy-mindedness has inevitably reduced the quality control that existed some years ago. It was once common practice to submit preliminary fire protection plans and design criteria to the authorities, as well as to the insurance interests having jurisdiction, for their review, comments, and approval. Only after these plans were returned with the comments and stamp of approval did the final design begin. Today this process of coordination is lacking, resulting in claims, delays, and change orders during the installation phase.

National Fire Protection Association (NFPA)

The National Fire Protection Association is not of itself an authority having jurisdiction, because the NFPA does not approve design plans or conduct tests or inspections of system installations. The NFPA does not approve or certify equipment or materials, nor does the NFPA single out or approve testing laboratories that evaluate equipment, materials, or systems.

The authority, or the insurance organization that does have jurisdiction to approve plans and installations may, however, base approval of the plans and the installation on compliance with NFPA Fire Code requirements. As such, the NFPA is the authority, but only within the framework of the local code or the insurance carrier that has jurisdiction. Understanding the origin of the NFPA Fire Codes is important because it explains the reasons and significance of codes in fire protection today.

When automatic sprinklers were first developed beyond the experimental stage, and began to be installed in buildings, several pipe sizing methods were in use. Differences in rules produced installations of all sizes and configurations. There was no proven way to ascertain the effectiveness or reliability of these installations, since there was no clear track record. Insurance companies were providing premium savings for sprinklered buildings, yet with no established standard of installation, they could not establish a foundation of sprinkler installation reliability.

As a result, a group of representatives from insurance carriers met in Boston in 1865 to establish a uniform standard for sprinkler installations. Their purpose was to obtain the installation uniformity, system performance reliability, and efficiency needed to establish realistic sprinkler premium rates. This meeting resulted in the first standard based on the consensus of opinions of experienced personnel for the design and installation of automatic sprinkler systems. This first consensus led directly to the birth of the NFPA in 1896.

The small group that established the first technical committee has grown into over 175 NFPA technical committees, consisting of over 3500 volunteer committee members, offering a vast source of experience required to develop, monitor, and periodically update over 270 fire codes. These NFPA technical committees have jurisdiction over three basic NFPA documents: the *Standard*, the *Recommended Practice*, and the *Guide*. NFPA Standards mandate requirements with the use of the word "shall." These requirements are, in most cases, detailed and specific design and installation requirements. NFPA's *Recommended Practice* and *Guide* are exactly what the names imply, and the word "should" is used in these documents in lieu of the word "shall."

A note of caution: The majority of the NFPA Standards have a section at the back of the Standard called the *Appendix*. The Appendix is offered as part of the Standard for information only, and is not part of the body of the Standard. The words "should" and "may" appear in the Appendix. Since Appendix items are not part of the body of the Standard, they are not mandatory requirements, but are for information only, and are used to explain a section of the Standard.

According to the administrative authority of NFPA Technical Committees and the Fire Codes, all Standards, Recommended Practice, or Guides must be reviewed, updated, or reaffirmed in their present content by the Technical Committee in charge of the document every five years, and a new issue of the code is printed each year. It is very helpful to know that a vertical line beside a Section or paragraph or sentence in a Fire Code,

Standard, Recommended Practice, or Guide is an indication that this is new wording since the publication of the last document.

The numerous volumes of Fire Codes contain all the current Standards, Recommended Practices, and Guides, but the date of the current issue of the Fire Code volume is not necessarily the latest issue of the particular document being used. For example, the 1989 set of National Fire Codes, with "1989" printed on each volume, may contain Standards dated 1988 or 1987, etc. If the Standard, Recommended Practice, or Guide is being referenced by date in the specification, be sure to check the date of the specific document. This may sound like an unnecessary caution, but changes in an updated document, particularly a Standard, can cause serious problems if the date used in the specification does not coincide with the actual date of the document.

The NFPA Fire Code requirements effectively become the jurisdictional authority when they are used as a basis of acceptance or approval of a system design or installation by the proper authority or the insurance interest having jurisdiction. NFPA Fire Codes actually become law when referenced by number and date of issue in a building code, and can be considered as a "law" when referenced in a specification, by a fire marshal, government agency, local fire prevention bureau, fire department, health department, or any other similar authority or insurance carrier having jurisdiction.

The NFPA may reference other Standards that had some bearing on a particular Standard under consideration. NFPA references these other Standards in the body of their Standard. In such cases, it becomes mandatory to use referenced Standards in the design and installation of the system, if applicable to the design and installation of the particular system.

Insurance Service Office (ISO)

The Insurance Service Offices are supported by the many stock insurance companies that conduct business in the state. One of their many functions is to determine insurance rates for those insurance firms in the state that subscribe to the ISO. ISO offices serve every state except Texas, where insurance rates are established by a state-supported bureau rather than an insurance supported organization.

The ISO evaluates the fire protection provisions and facilities of municipalities in their state in consideration of factors such as water supplies and fire equipment, and grade the municipality to establish an insurance rate structure for the area. The grading of a municipality determines the insurance premium the insurance agents use. This is a very basic description of the ISO functions regarding fire protection, but for this text the relationship of the ISO as the authority having jurisdiction is the important consideration.

The ISO used to be "The Inspection Bureau," and when suppression system plans were submitted to the Inspection Bureau, they were thoroughly checked. Every pipe size, head location, head spacing, pipe hanger, etc., was checked, and if not in accordance with NFPA Fire Code requirements, the plan discrepancies were "red marked," returned to the preparer for correction, and required to be resubmitted to the Inspection Bureau. A fire protection contractor would not even think of commencing installation until the plans were stamped "approved" by the Inspection Bureau.

Some drastic administration changes within the Inspection Bureau resulted in a complete reversal of the approval process. The Inspection Bureau changed its name to the Insurance Office of—the name of the state where the office was located. Under the present organization, when a major insurance carrier is not involved, and the insurance agent, or agency, requires that the plans be submitted to the ISO for review, the plans will be returned with a stamp that reads: "Reviewed for Rate Making Purposes." Very seldom will there be a red mark on the plan or an indication that the system does not meet NFPA requirements. This stamp tells the whole story, that the plans were reviewed, and the system protection as indicated has established the premium the agent should propose to the client.

The important thing about the requirement that plans be submitted for review by the ISO is that it puts all the responsibility of a proper design meeting the requirements of the applicable NFPA Standard entirely on the designer. If no other agency or authority having jurisdiction is required to review the plans, the designer has the full responsibility to provide a system that meets all the requirements of the applicable NFPA Standard.

National Electric Code

The National Electric Code (NEC) warrants individual recognition because this NFPA Fire Code, Standard 70, is considered to be law throughout the United States. The NEC has mandatory requirements that affect all types of electrical installations, of which fire detection and alarm systems are only one facet of the diversified code requirements.

One section of the NEC contains the design criteria and material, equipment, and devices to be used in areas subject to potential flammable liquid vapor or dust explosions. There is also a section on the location of explosion-proof devices and equipment relative to a fueled aircraft in an aircraft hangar. The important feature of the NEC is that it must be considered in the evaluation of every detection and suppression system and every hazard condition or situation.

Insurance Carriers

There are many major insurance carriers. We will consider two of them, in order to convey a working knowledge of the importance of dealing with this authority.

Industrial Risk Insurers (IRI)

In 1975, the Factory Insurance Association and the Oil Insurance Association merged and formed the Industrial Risk Insurers. The old FIA was a familiar name in the fire protection industry, and at that time consisted of over one hundred of the largest and strongest stock fire insurance companies in the world. This merger not only strengthened, but also diversified, the organization. To the fire protection industry, the FIA merely changed its initials to IRI.

The home office of the IRI is in Hartford Connecticut. They maintain regional offices throughout the United States, plus field offices and field representatives in most major cities. The IRI maintains its own engineering and inspection departments to review installation plans and make inspections of properties of their policyholders. IRI also has its own underwriting department that provides the insurance rates for IRI clients, and handles loss claims for policyholders.

The IRI insurance policies cover a comprehensive range of protection, including property loss from fire, windstorm, explosion, hail, lightning, smoke, riot, sprinkler leakage, vandalism, etc. Another important policy that is offered by IRI is for "Business Interruption." This policy covers losses that are the result of operations interrupted by several unforeseen occurrences, one of which is fire. Business interruption can play an important part in an evaluation of a hazard or a system selection, and sometimes will justify the cost of a specific type of suppression system when the hazard itself is not of great physical value.

A piece of equipment might be worth only $10,000, which would not justify the expenditure of $50,000 for a special hazard protection system, yet the loss of this piece of equipment due to a fire could preclude use of the facility for several weeks while repairs are conducted, or a new piece of equipment ordered, and the lost production for this period could run much higher than the replacement cost. Business interruption insurance could be very costly if a key piece of equipment is not adequately protected.

IRI generally uses NFPA Fire Code requirements, but beware, IRI often supplements the NFPA requirements with their own, which may be much more stringent. It is always good policy when designing protection for a facility that falls under IRI jurisdiction, especially if the facility is to be classified as a "Highly Protected Risk" (HPR), to obtain the specific requirements and design criteria from IRI. The protection requirements for an HPR include not only sprinkler and/or special hazard equipment, but also take into account construction features.

The insurance rates charged by IRI are relatively low by comparison with other insurers. This is because of their high standards of construction and protection. Very few IRI client facilities are not fully protected with automatic sprinklers. In addition, IRI has an excellent inspection service that will maintain a constant check of their policyholders' facilities and the protection of these facilities.

Some facilities have such a high value that a private insurance company cannot carry the risk of a loss on their own, and IRI becomes a partner in the risk. When this occurs, the IRI requirements take precedence over all other requirements, including those of the NFPA. When a specification lists the insurance interests having jurisdiction on a project, and IRI is included, it is good engineering practice to work with the IRI office that carries the risk. Without the complete assurance that the IRI requirements are being incorporated into the project, the project may be subject to a substantial cost impact at some point in the design.

The following is an actual example that shows the importance of communicating with the IRI engineers when developing the design criteria for a project:

> An architectural/engineering firm had been commissioned to design a manufacturing and storage facility. The storage part of the facility involved rack storage of combustible liquids in plastic containers packaged in cardboard cartons. The racks were to be constructed to a height of 90', and the cartons were stored on the racks automatically by a computer-operated lift arrangement.

> The fire protection was to be designed by the mechanical engineers in the firm, and they based their design of the rack sprinklers on the NFPA Standard 231 C, which is the Standard for rack storage. NFPA 231 C mandated that the hydraulic calculations for the rack sprinklers include seven sprinklers on each of the top two lines in the racks, or a total of 14 sprinklers.

The design of the rack sprinklers was finalized, the hydraulic calculations completed, and piping sized based on the calculations so that the rack manufacturer would have some concept of piping size and weight. Cost estimates were prepared based on the design. It was only at this point that the engineers made the decision to review their design with the IRI. They had complied with all the requirements of NFPA 231 C, and felt confident that their design was adequate. They discovered that IRI had compiled a history of testing with rack storage of combustible liquids in plastic containers stored in cardboard cartons. Based on these tests, and because of the unusual height of the racks on this particular project, the IRI requirements called for 14 heads on each of the two top lines, or a total of 28 heads.

Briefly, this made necessary the preparation of new hydraulic calculations and subsequent resizing of the rack sprinkler piping, but this was not the worst cost impact. The size of the fire pumps for the project had been based on the original 231 C design criteria and the original calculations. When the additional heads were calculated, it was determined that the pumps were undersized. This entailed redesign of the pump, pipe sizing, and pump accessories. The cost estimates were revised to reflect these changes. It was, therefore, necessary to amend the original budget, which the client believed to be fixed and reliable.

Factory Mutual System

The Factory Mutuals were founded in 1835 based on the conclusion that large industrial and commercial companies needed to provide mutual fire insurance coverage and loss prevention. At the present time the Factory Mutual System consists of three insurance firms—Allendale Insurance, Arkwright Insurance Company, and Protection Mutual Insurance Company.

The Factory Mutual System is often referred to as "The Factory Mutual Family," and the system is, in the true sense of this analogy, a family unit which not only operates the business of insuring policyholders, but also has "family members" that provide many benefits to the fire protection industry.

The Factory Mutual (FM) Research Corporation is the member of the family that conducts research, and out of this research department have come equipment, devices, and systems that have benefited the entire fire protection industry and profession. One function of the FM Research Corporation is the testing of equipment, devices, and systems to determine if their reliability and efficiency would warrant the "FM Approval" label. The indication that a device, piece of equipment, or system is "FM Approved" provides the user with the confidence that it has been thoroughly tested and found to be worthy of use in a fire protection design.

The *Factory Mutual Approval Guide* is a publication that is constantly updated, and lists all the products, devices, equipment, and systems that are FM approved. The *Approval Guide* not only lists FM approved items, but also details the material and installation criteria. In some instances the detailed "specifications" outline the parameters of usage to remain within the FM Approval.

The Factory Mutual research activities are located on the East coast (Norwood, Massachusetts), and the facilities are staffed by experts in many fields, using the finest testing equipment and facilities to be found anywhere in the world. Factory Mutual Research makes these facilities and laboratories available to private companies and organizations on a contract basis.

Since FM policies can cover a multitude of loss coverages, their testing and requirements go far beyond fire protection. FM requirements include windstorm and hail, explosions, vandalism, sprinkler leakage, lightning, and loss or damage from furnace malfunction. FM publishes a great deal of literature on requirements for protecting and preventing losses from these hazards, but one publication in particular should be mentioned. That is the FM publication on fire resistance of various construction elements. This resource is filled with easy to understand details and descriptions that inform the user of what must go into a wall, ceiling, roof, or floor to provide the required fire resistance rating of one hour, two hour, three hour, etc.

Factory Mutual has always limited their insurance coverage to highly protected risks, and their requirements reflect this environment. A very important member of the FM family is the Factory Mutual Engineering Department. These departments are staffed by experienced professional fire protection engineers who review the design and installation plans on all FM projects.

NFPA has their Fire Codes, and Factory Mutual publishes their requirements in the Factory Mutual Data Sheets. Make no mistake, if the project is a Factory Mutual client, use the Factory Mutual Data Sheets to obtain the design requirements. Some Factory Mutual Data Sheets will use an NFPA Standard word-for-word, but FM will always add in bold print any additions to, or changes from, the NFPA requirements.

In addition, FM has a staff of experienced, professional, field personnel who conduct the testing of new systems; these field personnel use the testing requirements of Factory Mutual. When specifying the testing of the systems, be sure that the specification section is based on FM requirements. This aspect of a project also can have a cost impact, because the installation contractor may be required to furnish additional labor, material, and equipment to satisfy the test requirements.

Architects and engineers who have a client that is, or will be, insured by Factory Mutual should always arrange meetings with Factory Mutual personnel at the start of the project.

Factory Mutual offers the FM Data Sheets to indicate their requirements, but every FM project protection system is tailor-made for the specific hazard and/or facility. FM may require other design features not covered in the Data Sheets. FM fire insurance premiums are very low because they require the most effective and reliable protection that is available to protect their risk.

In summary, do not try to outguess the protection requirements on an FM insured risk, and do not look at the Data Sheets as the last word in requirements for the design. The only sure way to provide the client with a Factory Mutual approved system is an exchange of communication with FM personnel.

These meetings with FM are not being recommended just for fire protection discussions. On new building construction design, they must cover all aspects of loss prevention construction, and architects and engineers of all disciplines on the design team should participate.

If FM is brought in late in the design process, the architect and engineer may find that they have overdesigned or underdesigned the facility, including the fire protection.

Building Codes

Building Codes are the product of a jurisdictional authority having the power of the law. This power is administered by the building code officials of a Building Code Department.

Municipal building codes, and those codes that have been adopted to cover an entire state, provide the requirements to establish an occupancy classification, and many of the construction features that apply to the various occupancy classifications.

The major problem for the architect, engineer, and fire protection engineer is that there are many sources of building codes, and each code has its own set of requirements and mandates. Three of the many building codes that will most frequently be used are: The Uniform Building Code (UBC), produced by the International Conference of Building Officials (ICBO), the BOCA Building Code (BOCA), produced by the Building Officials and Code Administrators (BOCA), and the Southern Building Code (SBC), produced by the Southern Building Code Congress.

The important thing to remember in dealing with building codes is the fact that they are written to cover a multitude of different types of facilities and construction types. Every project is different, with different occupancies, and different types and numbers of occupants. At times it becomes extremely difficult to fix an occupancy classification for a particular facility, and interpolating the various requirements of the code to provide compliance with the code requirements for the specific facility may seem like attempting to fit a square peg into a round hole.

It has been said that 80% of building codes relate to property protection and life safety, and therefore it is important to evaluate the requirements of the code having jurisdiction prior to a determination of how much fire protection and what type is required. In the majority of building codes, the occupancy classification will determine whether on not fire protection is required, or what type of protection is required.

An important point in dealing with building codes is best explained by a simple example:

> In evaluating the building code having jurisdiction, it is determined that based on the occupancy classification and building configuration, smoke and heat venting at the roof elevation is required. In the section of the code pertaining to smoke and heat venting, specific requirements are presented regarding spacing of the vents and the number of square feet of venting area required. In this case, the building owner attempted to justify the use of power venting to substitute for some of the mechanical vents. The owner, in effect, asked the Board for permission to provide an equivalent amount of venting by alternate means.

The fact is, the designer is violating the code and the law if the decision is made to reduce the number or increase the spacing of vents, if this will result in a reduction of the venting area.

There have been instances where the "authority" to modify a code requirement consisted of the oral approval of a building inspector, or even a building official. By accepting this "authority," the architect and/or engineer is not only violating the law, but opening the door to liability problems.

If there is some requirement in the code that presents a tremendous hardship from a design, operational, or cost standpoint, and the architect/engineer feels that there is ample justification to seek a variance to the code requirement, the only recourse is to apply to the building code department for a variance. Application for a variance is really an appeal to appear before a Board of Appeals, present the justifications, and convince the Board that the variance will not lessen life safety or property protection. This justification may take the form of alternate methods of providing an equal measure of life safety and property protection—alternate methods that may not comply with the letter of the code, but do satisfy the intent.

The Board is made up of architects, engineers, and contractors who are not employed by the Building Code Department. The Fire Chief and/or a Building Code Official will be able to present their recommendations to the Board, but will not have a voting privilege.

It is important to keep in mind that a Board of Appeals does not have the power to change or delete a code requirement, any more than a review board of this type has the power to change or delete a law, regardless of the circumstances. The Board of Appeals only determines the suitability of the alternate methods of construction or materials, or provides a reasonable interpretation of a code requirement.

It is stated in most of the Building Codes that the Building Official has the authorization to enforce all the provisions of the code, with the power of a law enforcement officer. The Building Official can and does exercise this power by refusing to grant an Occupancy Permit if the facility does not meet the code requirements. This is why it is so important to design and construct a facility that meets all the requirements of the code, because there is hardly any other cost impact that affects a client more severely than to have his facility completed and not be able to obtain an occupancy permit after the building inspector finds a serious code violation, or a deviation from the plans approved by the Building Code Department.

The architect, engineer, and/or client must submit plans, specifications, and computations to the Building Department in order to obtain a permit. This permit may be for construction of a part of the structure, or it may be a permit for all the construction features. The cost of the permit is based on the overall project cost.

From a fire protection standpoint, the occupancy classification may or may not determine the fire protection required. Some building configurations mandate a requirement for a suppression system regardless of the occupancy classification. It becomes necessary to carefully evaluate the code requirements relative to the architectural design of the structure.

Understanding occupancy classification is important. Take the following example:

> A building classified as an office occupancy, and having the size, height, and building materials under this classification did not require the installation of a sprinkler system. Once the project was well-developed and the budgets established, it was discovered that one section of this same building code stated that all occupancy classifications with a basement or story that exceeded 1500 square feet that did not have window openings of a certain size, and spacing, located in walls that were entirely above the ground level, were required to be protected with automatic sprinklers.

> The facility in the example had a basement that exceeded the 1500 square feet, and being below grade, did not have any windows. It was, therefore, necessary to install a sprinkler system. Windows are required for access by the fire department to facilitate firefighting and rescue from the exterior of the building; therefore, where windows are not possible (below grade), a sprinkler system and detection/alarm may be deemed necessary.

Most building codes are updated periodically, and it is important to determine the date of the code being applied. This is especially true when writing a specification. Some building code departments may not use the current issue, and also if the project has a lengthy construction period, a new code issue may have required changes or additions.

Most building codes have what are commonly termed "trade-offs." These trade-offs involve the allowance of construction features if a sprinkler system is installed. In other words, if an occupancy classification and type of construction material limits the allowed square footage or height, either may be increased if a sprinkler system is added to the building design. Some of these trade-offs can be very cost-effective, but they should be studied carefully because most building codes state that the sprinkler system cannot be used to accomplish a construction trade-off if such a system is already required for the project.

Most building codes provide Fire Codes that have the same legal authority as the building code used by the municipality. The Fire Chief or Fire Prevention Bureau of the municipality usually assumes jurisdiction over the Fire Code. Fire Codes in general cover special hazards such as combustible and flammable liquids, combustible fibers, spray finishing, explosives, and gases, but they also encompass fire prevention and a multitude of processes that go beyond the building code requirements.

The fire department may have the authority to issue permits for specially hazardous processes or storage of hazardous materials, and may have the power of law when fire inspectors perform inspections of these facilities.

It is always extremely important to conduct a careful evaluation of the Fire Code as well as the Building Code. In most cases the Fire Code will follow the requirements of the Building Code, but there are many requirements that may only be found in the Fire Code, and the architect, engineer, and contractor must, by law, comply with both the Building Code and Fire Code when both have been adopted by the municipality.

Fire Marshal and Fire Prevention Bureau

Discussions of the Fire Marshal and the Fire Prevention Bureau as the authority having jurisdiction must be integrated, because in many cases this authority is one and the same. This section follows the dissertation on the Fire Code because the Fire Code is a basic regulatory tool of the Fire Marshal and the Fire Prevention Bureau.

Not every state has a Fire Marshal, but all do have an authority with the same responsibilities and power of law regarding fire protection and prevention, whether or not this individual carries the title of Fire Marshal.

In some states the office of the State Fire Marshal is a separate department of state government, while in others, the State Fire Marshal's office is under the supervision of the Department of Insurance. In another example, the State Fire Marshal's office is a separate government department reporting directly to the Governor. Regardless, it is important to understand that this authority in every state demands the attention of the architect, engineer, and contractor if projects within the parameters of state jurisdiction for fire safety are to be approved by all authorities having jurisdiction.

The duties and authority of a fire marshal are, in general, to enforce the law regarding storage, sale and use of combustibles and explosives; the means and adequacy of egress for institutions and places of assembly; the investigation of fires as to their causes, and; the suppression and investigation of arson. In addition, most fire marshals are responsible for performing fire prevention inspections.

These duties vary from state to state, but it is safe to say that if a project involves a public building, and especially if the public building is a place of assembly for large numbers of people, or an institutional facility, such as a hospital or nursing home, or for the use or storage of flammable, combustible liquids, or explosives, the fire marshal's office will be involved. The project approval must include the Fire Marshal's office or the agency acting under the jurisdiction of the Fire Marshal.

These state agencies usually have jurisdictional parameters beyond that of municipal or county organizations, but the authority to perform as agents of the state on a local, municipal, or county level can be granted by the state. This delegated authority covers specified areas such as inspection, investigations, and enforcement.

On this local level, the municipality or county can delegate authority, by law, to a Fire Chief, or in some instances, to the head of the fire department or fire district. This individual is usually given the title of Fire Marshal as well as the authority to delegate the responsibility of carrying out the mandates regarding fire prevention and protection to an individual (customarily a Fire Chief) or, depending on the size of the fire department, to a division within the fire department.

Many states with large municipalities or highly organized counties or fire districts within the state have a state fire marshal or agency with similar authority, and each large municipality, county, or fire district has a local fire marshal and/or a Fire Prevention Bureau under the direction of the local fire marshal.

For example:

> In one state, a fire marshal is responsible for enforcing all state laws regarding arson, and for providing state regulations and requirements for flammable liquid use and storage, as well as for schools, hospitals, hotels, etc. Two large municipalities within the state have local fire marshals who are in charge of the municipal Fire Prevention Bureau.

> A fire protection project in one of these municipalities under the jurisdiction of the local Fire Marshal (and therefore under the jurisdiction of the Fire Prevention Bureau) must obtain approval of plans from this agency. The Fire Prevention Bureau may even take precedence over the local Building Code plan approval, especially if the project falls directly under the guidelines of the Fire Code requirements.

When it comes to fire protection and/or life safety, the authority and requirements of the local fire marshal acting as head of the Fire Prevention Bureau, backed up by the enforcement of law—even outside the scope and letter of the Code—take precedence when it comes to plan or installation approval.

The architect, engineer, and contractor are advised to become familiar with the state, county, fire district, and local authorities having jurisdiction within the parameters of the specific project, the state, and the local area of the project within the state.

Keep in mind that the fire inspector from the Fire Marshal's office, local or state, or the Fire Prevention Bureau, can cite installation deficiencies and violations that do not appear in the Fire Code, although these are usually of a minor nature. If the installation has not deviated from the plans approved by the Building Code Department and the Fire Prevention Bureau, these inspection citations can usually be corrected inexpensively and quickly.

Other Authorities Having Jurisdiction

Federal authority having jurisdiction, and the requirements of federal agencies, must be considered in a different environment, as the federal agencies and organizations have, for the most part, their own set of requirements and regulations.

The Department of Defense provides a military handbook, entitled "Fire Protection for Facilities—Engineering, Design, and Construction."

This military handbook details the types of construction and lists occupancy classifications similar to the building codes previously described. In addition, and again similar to the building codes, the handbook details allowed areas and heights of structures, interior finishes, air handling systems, and suppression and detection systems.

Actually, the handbook is a building code, a life safety code, a fire code, and the equivalent of the NFPA Fire Codes—but on a government project, the military handbook takes precedence over the requirements of the NFPA except where the NFPA requirements are specifically spelled out as the design criteria, or a specific NFPA standard is stated to be the required standard to follow.

A similar publication is produced by the Air Force. This is the "Air Force Design Manual Criteria and Standards for Air Force Construction", further defined as "Facility Design and Construction." The General Services Administration, the Department of the Interior, and the Department of

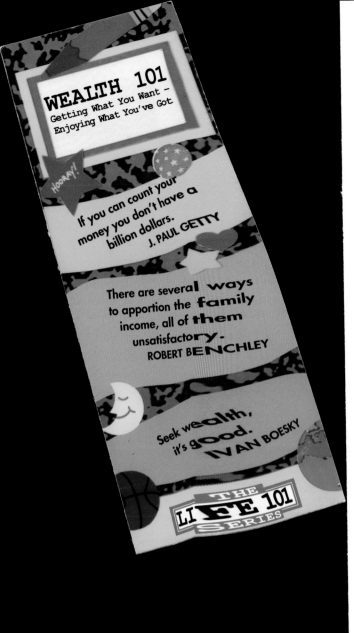

Transportation also publish standards that cover their specific area of jurisdiction and authority. The Environmental Protection Agency (EPA) also publishes standards, and the EPA can greatly impact fire protection projects. An example is a foam suppression system. When a foam system is installed in an aircraft hangar, it is required in the majority of cases that the foam discharge be retained and not allowed to enter a stream or flow onto the ground. In some instances, this requires huge tanks or concrete catch basins to retain the foam discharge. AFFF (aqueous film forming foam) is biodegradable, but only if it is passed through a waste treatment plant at a slow rate, so the retention device must hold the discharge until it can be pumped into the waste treatment system.

Water Department

It may seem strange to include the water department under authorities having jurisdiction, but this is an agency that must be given consideration. Two very important items fall under the jurisdiction of the water department, and both can have a definite cost impact on a project—a major cost impact if the project has already been installed.

One is metering. Some water departments require that fire suppression systems that take supply from a city water main have a dedicated meter on the supply line. Two major questions must be answered. First, does the water department in the area where the project is located require the metering of fire suppression systems? If the answer is yes, the second question is: do they only require the metering of small flows, and not the metering of fire flows when the system is operating?

If the requirement is only metering of small flows, the specifications will require a UL-listed detector check meter. Small flows are metered through a bypass with flow through the check restricted by a weighted check valve.

In some cases the water department will require metering of both small flows and fire flows. The specification must now include a full flow fire meter. These are large devices, UL-listed, with an OS&Y valve on either side of the meter so it can be removed for repair or replacement.

With the detector check valve, the meter is in a small line bypass around the check valve, and with two shut-off valves on either side of the meter, the meter can be removed for replacement or repair without shutting down the fire line supply to the suppression system.

With the full flow fire meter, the two control valves are installed so that the entire meter can be removed for replacement or repair. When this occurs, the fire line supply to the suppression system is shut down. To avoid losing supply to the suppression system, there is usually a bypass around the meter with a normally closed control valve. If the full flow meter were removed, the bypass could supply water to the suppression systems to keep the fire protection in service.

Metering, if required, is an important specification item, and a quite expensive equipment and installation item, especially if it involves the full flow fire meter. It is well worth the time and effort to establish definite design requirements with the water department.

The second area where the water department is the authority having jurisdiction is backflow prevention. Most U.S. cities are now requiring backflow prevention on fire suppression systems that take supply from a potable water supply.

Similar to the metering requirements of the water department, each water department that requires backflow preventer devices has specific requirements for a double check backflow preventer or a reduced pressure backflow preventer.

Always determine if backflow prevention is required in the water district or water department where the project is located. It is then extremely important to verify which type of backflow preventer is required. The following is an actual example of the problems that can occur when all of the information is not obtained.

It was learned that the water district where the project was located required backflow prevention on automatic sprinkler systems taking supply from a city water main, but no one had obtained all of the water district requirements.

The specification writer specified a double check backflow preventer, and the designer of the preliminary plans for bidding designed around a double check backflow preventer. The successful bidder included the cost of a double check backflow preventer in his estimate, and proceeded to install his device.

After the installation was complete and ready for inspection, the code department, which was fully aware of the water district requirements, denied the occupancy permit. It was then discovered that the water district requirements allowed a double check backflow preventer on "standard" sprinkler systems, but the requirements mandated a reduced pressure backflow preventer when the sprinkler system was supplied through a fire pump, and this particular project sprinkler system had a fire pump.

The cost impact, and the legal problems concerning who was to pay for the change, were overshadowed by the delay the client suffered because he was denied an occupancy permit for several weeks.

Summary

The purpose of this chapter has been to impress upon the architect, engineer, specification writer, and estimator the critical importance of defining who is really the "authority having jurisdiction" over the project, and not leaving a specification with generic wording encompassing a multitude of authorities so no one will be missed. Defining exactly what agency, individual, organization, or insurance carrier is the authority having jurisdiction on a specific project may require a time-consuming evaluation, but in the final analysis, it will prove to be worth every effort and every hour.

Many a project has suffered a severe cost impact, time lost, and schedule destroyed impact, not to mention damaged "PR" as a result of a "quick and dirty" specification, where every conceivable authority having jurisdiction is listed without having determined if some on the list could come back to haunt the project, or someone was omitted from the list who will appear later on to devastate the installation approval.

CHAPTER ELEVEN

Suppression System Remodeling

This chapter is not a "how-to," or a "do-it-yourself" set of guidelines on the design and engineering of suppression and detection systems, but rather is intended to provide some information, cautions and, hopefully, helpful insight into some of the idiosyncrasies of the fire protection profession, especially those relating to remodeled and revamped systems. In this way, the designer/engineer, or contractor will be better prepared when these situations arise, and can tailor the solution to the individual problem.

The cost of new construction, the high cost of land on which to expand or build a new facility, and the time demands of new construction have made remodeling and revamping existing buildings a major operation in every type of business, industry, or marketing enterprise. The majority of these projects involve either the remodeling or revamping of an existing suppression or detection system, or installing a new suppression or detection system where none was previously required.

Remodeling Existing Systems in Occupied Buildings

Remodeling and renovation projects in buildings with an existing sprinkler system usually involve the remodeling or renovation of the sprinkler system along with the construction features of the facility. If the renovation or remodeling is to take place in a building where the operations of the facility are to continue uninterrupted (other than the inconvenience of the work area), the fire protection system requires special attention from a cost, specification, design, and installation standpoint.

Those in charge of fire protection must exert every professional effort to minimize the inconvenience and interruption of business, production, and operations during system retrofit or remodel, and at the same time maintain the minimum interruption to protection. Not an easy task to accomplish, and unless carefully planned to receive a high priority in every aspect of the project, it will not happen. The estimator must be well aware of this planning criteria, and the client's desires and options must be thoroughly assessed, understood, and integrated into the planning. Otherwise, the unanticipated cost impact could be quite severe.

The following example of the installation of a sprinkler system in an existing and operating office building illustrates this point:

The specification and the cost estimate reflected the fact that a thorough evaluation of the client's desires and concerns were not investigated. When the installation crew arrived, a day was wasted attempting to find a "lay down" storage area for pipe, material, and equipment, and a work area—and a day for a full crew is not a cheap venture. Then came the first contact of the installation crew with the client. At this meeting, the crew informed the client how they intended to proceed with the installation and learned that no work was to be performed above the office employees because it would not only interfere with their work, but also (due to the liability concerns) cause a tool or material to fall and cause injuries. This led to hit-and-miss installation in unoccupied areas during the day, with a second crew working after hours in the normally occupied areas.

The additional cost of the second crew was not too severe, since some extra compensation was agreed upon in advance, but the fact that the day crew was not able to install the piping with systematic productivity was costly. They were forced to install a few heads and piping here and there without being able to tie the system together in a normal procedure.

One additional unexpected cost item which the contractor was lucky enough to at least share with the client, was that of hiring a security guard to accompany the after-hours crew, because some of the offices were handling legal and sensitive material.

Had the initial investigation of the project been thorough, and included the client's concerns, planning the work areas for the day crew and designing the system to accommodate their work areas would have resulted in a profitable project and a satisfied client. The blame for this project fiasco which verified "Murphy's Law," could be equally shared between the architectural/engineering firm that prepared the specifications and the contractors who bid the project, for not questioning the specification before the project was bid or requesting a pre-bid meeting to discuss the possible problems.

Additions to Existing Sprinkler Systems

If an addition to an existing sprinkler system is to be supplied from an existing sprinkler system piping, it is important to be sure that the existing system can be shut down long enough to make the connection, or cut in during regular hours when the facility is operating. Shutting down an existing system to make a connection may eliminate protection over a wide area of the facility, and this may not be allowed by the client or the insurance carrier.

If an architect/engineer is preparing plans and specifications for a project, the existing system shutdown (in order to make the connection to supply the new work) should be discussed with the client and the insurance carrier. Once an agreement has been reached, the details should become a prominent section of the specification.

The same scenario is true if a fire protection contractor is negotiating for the work and an A/E is not involved. The agreement should be a part of the contractor's contract so there is no misunderstanding that will add unexpected cost to the project, or subject anyone to potential liability.

This written agreement should have a clause to the effect that the contractor will notify the client, and possibly the insurer, at least five days prior to the shutdown and connection. This is important regardless of any agreement for the shutdown because it gives the owner sufficient time to notify any security personnel to arrange their schedule to be on the alert for fires during the absence of automatic protection, or the assigning of a person as a "fire watch" during the down time.

It is recommended that, if at all possible, the new work be installed back to the point of connection to the existing system. Before the connection is made, if possible, blank off and hydrostatically test the new piping. This will eliminate the need to test the existing system and the new piping after the connection is completed.

It is usually necessary to make this connection on a continuous basis, and once the connection work starts, it should go to completion. Be sure any overtime hours that become necessary are agreed upon before the work starts, because "Murphy's Law" always comes into play during, and in spite of, the best laid plans of the engineer, designer, or installation crew.

On the day agreed upon for the shutdown and connection, all material for the connection should be on the job site, with a few extra pieces of material on hand if at all possible.

New work must be tested prior to making the connection to the existing system. Many times, an addition to an existing system does not consist of simply connecting a supply pipe to the existing system. The rearranging of several heads with new piping, or the addition of new heads on existing or new piping does not provide the luxury of blanking off a single supply pipe and testing the new work before it is connected to the existing system. Herein lies the important consideration.

The architect/engineer preparing the specifications for an addition to an existing sprinkler system, or the remodeling of an existing system, should always include a clause to protect both the contractor and the client. The following is merely a suggested wording of such a clause:

> "If it is not possible to test the new sprinkler work without testing the existing piping, any leaks or damage that occur within the existing system piping as a result of the hydrostatic test will be called to the attention of the owner or client, a proposed cost of repair submitted to the owner or client, but the corrective work will not be performed without a written agreement from the owner or client. Any work on the existing system as a result of the hydrostatic test of the new work is outside the scope of this contract."

Do not consider the above to be the legal or correct words to use in a specification or contract, but rather the intent expressed by this paragraph should be included in the specification or contract in order to save time delays, contract problems, and possibly a cost impact on the project.

Suspended Ceiling Remodeling

One of the most common remodeling projects involves installing a suspended ceiling in an area where this type of ceiling did not exist before, and where existing sprinkler piping is located at the original ceiling elevation with upright sprinkler heads under the building construction.

Again, it is important to repeat the statement made at the beginning of the chapter: Our purpose is not to provide the reader with a "how-to-do," or a "do-it-yourself" guide to design or installation, but rather to present some of the problems involved in this work, and offer some suggestions to stimulate the designer, engineer, or contractor to come up with their own innovative solutions to a specific situation.

When a new suspended ceiling is being proposed in a sprinklered area with sprinkler piping using upright sprinkler heads installed above the proposed ceiling, it is the first order of business to obtain a reflected ceiling plan of the new ceiling. The reflected ceiling plan is a plan that would reflect the ceiling if the floor were a mirror. Therefore, this plan indicates the arrangement of the ceiling tile, recessed lights, HVAC diffusers, speakers, and anything that will be installed in, or on, the new suspended ceiling.

Using the reflected ceiling plan, spot the location of the sprinkler heads. This is not always an easy task when one is attempting to get a symmetrical head design, maintain the required head spacing and coverage, and miss the lights and all the other obstructions on the ceiling.

If possible, obtain a plan of the existing system. If none is available, it will be necessary to survey the area and spot sprinkler piping, piping size, and existing upright head locations. If the existing system has the same number of existing heads as you have spotted on the reflected ceiling plan, proceed as follows:

On a plan showing the existing system lines and heads, spot the head location of the new heads that will be installed below the new suspended ceiling. Considering the existing head removed, show a 1" arm over to the new pendent head below the ceiling from the existing upright head.

This is accomplished—and should be shown with a detail similar to Figure 11.1—by removing the existing upright head, installing a 1/2" by short nipple with a 1/2" by 1" elbow on top of the 1/2" nipple. The 1" arm over to the new head location will have a 1" elbow to accommodate a 1" drop nipple to the new pendent head. The 1" drop nipple will have a 1" × 1/2" reducing coupling on the bottom to receive the new pendent head below the ceiling.

With this example project, if the number of upright heads on the existing piping is equal to, or less than, the number of heads below the new ceiling, replacing the upright with pendent heads will not necessitate replacing or resizing existing line or supply piping.

There can be an extra expense regardless of the number of heads that must be changed from upright to pendent. The additional man-hours include the time required to drain the system down before beginning work, and to restore the system to service when the work is completed. If the entire existing system covers a large area, and the new suspended ceiling is a considerable part of this protected area, the time required to drain the system down every morning and restore it to service every night can be considerable, and should be included on the estimate sheet.

On the majority of projects of this type—in fact on every remodeling project—this added expense of leaving the system in service whenever the installation crew is not actively engaged in the work—nights, weekends, holidays, etc.—can be considerable. This additional cost not only involves the time required for system draining and restoring to service, but also for plugging or capping off unfinished piping at the end of an installation crew shift before restoring the system to service.

Return bend arrangement.

Nipple and reducing elbow supplying
sprinkler below ceiling.

Sprinklers in concealed space and below
ceiling.

Suspended Ceiling Sprinkler Arrangements
(courtesy of the National Fire Protection Association)

Figure 11.1

The usual procedure for remodeling both this type of sprinkler system and new projects, where a new suspended ceiling is involved, is to make the installation before the suspended ceiling tiles are installed. If the ceiling grids, or ceiling tile supports, have not been installed when the sprinkler work is performed, it is necessary to establish the ceiling height of the suspended ceiling.

Having the ceiling grids installed, or knowing the suspended ceiling height, the usual procedure is to install the 1″ drop nipples to the future pendent heads cut to a length that will allow them to extend 4″ to 6″ below the ceiling grid or suspended ceiling elevation.

When the ceiling tiles are installed, and the 1″ drop nipple protrudes through a hole cut in the ceiling tile, the 1″ drop nipple is marked at the ceiling line, backed out, and cut and threaded for the 1″ × 1/2″ reducing coupling to be installed on the drop nipple.

When the drop nipple is re-installed, and the pendent head screwed into the 1/2″ outlet of the 1″ × 1/2″ reducing coupling, the pendent head protrudes below the ceiling at the ceiling tile elevation. An escutcheon, or ceiling plate, is installed to cover the ceiling tile hole, and installation is complete.

With careful workmanship in the marking and cutting of each drop nipple, and use of a shallow escutcheon, the sprinkler head will hardly be noticeable on the ceiling. If each drop nipple is not marked and cut carefully, not every head will be the same distance below the ceiling tile, and a shallow escutcheon cannot be used because it would hang down below the ceiling tile and reveal the hole. In this case, the contractor will use a deep escutcheon that is adjustable by being constructed in two parts that telescope into each other, allowing adjustment of the head elevation to accommodate the drop nipple length.

If appearance is important, and it should be with a suspended ceiling project, mandate in the specification that the escutcheons be one solid piece construction, and not over 1/2″ to 3/4″ in depth. This may seem like a minor point, but if the specification does not definitely dictate the type of escutcheon to be used, the installation may consist of pendent heads installed some 4-6″ below the ceiling. If appearance is not a major consideration, and the adjustable escutcheons are permitted, the labor cost may be slightly lower because the contractor may not have to cut every drop nipple. Stock, off the shelf nipples, can be used even if they are an inch or two longer or shorter than the required length.

Installing Pendent Heads in the Center of the Ceiling

Some suspended ceilings are installed in areas where appearance is extremely important. To maintain a symmetrical pattern of pendent heads on the ceiling, the heads must be placed in the center of the ceiling tiles or in a 2′ x 4′ ceiling tile, 2′ from the ends, and 1′ from the sides.

This arrangement must be clearly defined in the specification because it will add additional labor, and possibly additional material cost to the contractor's estimate. This additional contractor's cost will be incurred by the designer who may have difficulty in obtaining a symmetrical head pattern on the reflected ceiling plan by having to hit the center line of the ceiling tiles.

Placing the heads in this location may also be more costly because additional heads will have to be installed, and the usual method of installing these heads in this precise location may require what is commonly referred to as a "gooseneck," an up-over-and-down arrangement that necessitates additional material and labor.

The gooseneck arrangement consists of a short 1" nipple installed in the outlet of a tee on the sprinkler line, an outlet that is "looking up". An elbow is installed on the top of this short nipple, and a short, usually 6", horizontal arm over will position the drop nipple to the pendent head over the center of the ceiling tile. The advantage of this up-over-and-down configuration is the ability to "swing" the arm over slightly to position the drop nipple directly over the hole cut in the center line of the ceiling tile.

Why not drop directly out of the tees on the sprinkler line to supply the pendent heads? A sprinkler line may not be installed perfectly straight, and the heads would not, therefore, drop precisely in the center of the tiles. Or, the sprinkler pipe lengths in the sprinkler line may not be cut evenly, and the heads would not fall directly in the center of the tiles. This does not indicate poor workmanship, but it is easy to pick up 1/4" in pipe length on a sprinkler line, depending on how tight a fitting is made up, and if there are eight or more pieces of pipe between the tees, the drops will be 2" off. The up-over-and-down provides some adjustment flexibility to correct these differences. Note that when nonpotable water is used for the sprinkler system, this up-over-and-down arrangement is required for pendent heads on concealed piping.

Arm-Over Hangers

On projects with concealed piping above the ceiling and pendent heads below the ceiling, be sure to provide support for the arm-overs to the drop nipples that supply the pendent heads. This is especially important on the previously described remodel projects when the 1/2" nipple is used. The hanger is important so that the weight of the arm-over, regardless of the length, and the drop nipple and head will not be supported on the rather fragile 1/2" nipple. Hangers are usually not required on new sprinkler installations where the gooseneck is used to hit the center of the ceiling tiles, because the nipple out of the tee is 1" pipe, and the arm-over is kept as short as possible. Even on new installations, if the arm-overs are fairly long, but not long enough to require a hanger, it is good practice to provide a hanger to not only support the drop nipple and pendent head, but also to accommodate the upward thrust of the head when it discharges.

Other Sprinkler System Remodeling for Suspended Ceilings

Another remodel project might be an existing sprinkler system with concealed piping and pendent heads where new or relocated partitions would require relocation of heads and/or the addition of heads below the ceiling. To keep the cost to a minimum, it may be possible to relocate heads or add heads without disturbing the sprinkler lines.

Again, the first step is to establish the location of the heads on the remodeled reflected ceiling plan, and then by removing the 1" drop nipples, insert a short 1" nipple in the 1" tee outlet. This 1" nipple, with an elbow attached, can accept a 1" horizontal arm-over to the new head location.

If the existing sprinkler line pipe sizing is pipe schedule and will accommodate an additional head, attach a tee instead of an elbow on the 1" nipple, and use horizontal arm-overs in two directions. Sometimes the pipe sizing on a pipe schedule system will accommodate additional heads because the remodeling has eliminated a head or two.

If the existing system is hydraulically calculated, additional heads cannot be arbitrarily added, nor can additional fittings be used, without recalculating the system to be certain that the demand caused by the additional heads and/or fittings has not exceeded the supply.

Existing Dry System Remodeling

The previous discussion of remodeling of existing sprinkler systems below a suspended ceiling has been limited to wet pipe sprinkler systems. In dry systems with concealed piping above a suspended ceiling, there are no 1" drop nipples to pendent heads; all the pendent heads are UL-listed dry pendent heads. Dry pendent heads come complete with their own one-piece drop nipple containing a mechanism that prevents water from entering the drop nipple until the pendent head on the attached drop nipple fuses.

When it is necessary to convert an existing dry pipe system to a concealed pipe system with pendent heads below the suspended ceiling, it is necessary to remodel the sprinkler lines supplying the upright heads. This usually entails removing the lines and changing the existing 1/2" outlet tees to 1" outlet tees, and reinstalling the lines with the tee outlet looking down. The dry pendent heads are then installed in these 1" outlet tees.

With the permission of the authority having jurisdiction, which in this case is normally the insurance interest, it is possible to obtain approval to install pendent heads on regular 1" drop nipples, or the 1/2" up-over-and-down arrangement, when both the concealed piping and the area below the suspended ceiling are heated areas.

Delayed Installation of Suspended Ceiling

The best example of a building that will eventually have a suspended ceiling, but does not have this feature at the time of construction, is the new office building where space has not yet been leased, and therefore has yet to be partitioned. Another example is a new shopping center where spaces for stores have not been leased. A third typical example is the existing building that is being remodeled, but space allocations are not yet determined.

In order to keep the drop nipple length to a minimum, when installing the sprinkler piping, the future suspended ceiling elevation is established and the yet-to-be concealed piping is installed as close to the proposed ceiling elevation as possible.

This arrangement may place the sprinkler lines two feet or more below the structural ceiling. Often, the insurance interest, building code official, fire marshal, or fire inspector will require sprinkler protection when construction is complete, and heads installed in the yet-to-be concealed piping would, therefore, be far below the elevation required. This requirement for sprinkler protection, well in advance of the suspended ceiling installation, is unavoidable if the area is occupied by people, equipment, or storage.

In this case a 1″ riser nipple can be installed in the sprinkler line tees with a 1″ × 1/2″ reducing coupling on top and an upright head screwed into the 1/2″ outlet of the reducing coupling. This will place the sprinkler heads at the required distance below the ceiling structure, and provide sprinkler protection until the ceiling is installed. Once the ceiling is ready to be installed, the 1″ riser nipple can be removed, a short nipple installed in the 1″ tee outlet with an elbow on top, and the drop nipple to the new pendent head located on the elbow on the end of an arm-over to the pendent head location.

It is obvious that this arrangement involves additional cost which must be considered in the cost estimate, but it does provide the flexibility to arm-over to head locations in the new suspended ceiling, and the flexibility to adjust head locations with arm-overs for partitions that were not planned when the basic system lines were installed.

One last cost consideration is to be sure to oversize the basic lines in these areas to accommodate multiple partitions or a ceiling grid for which it is difficult to symmetrically design a head pattern. Usually, the number and possible arrangement of future partitions can be anticipated in sizing the lines, but a miscalculation can mean installing new lines and/or new supply piping to supply the number of heads in the finished area.

Valve Supervision with Remodeling Projects

In most of the remodeling projects previously described, the shutting down of a sprinkler system was necessary—either during installation of the remodeled system, or during the time required for a cut into the existing system to obtain supply for the new heads in an addition or a remodeled area.

A study of NFPA or FM records indicates that the majority of fires in sprinklered buildings that caused serious damage because the sprinklers did not control or extinguish the combustion, were the result of a closed sprinkler system water supply valve. Case study after case study of these fires indicates that the valve was closed for work on the system and never reopened. Most of the valves left in the closed position (after work on the system was completed) can be blamed on human negligence and carelessness, but some closed valves were left in that position overnight or over a weekend because the system work was not complete.

If the water supply control valves have electric tamper switches, closing the valve will activate a trouble alarm, both audible and visible on a control panel. The audible alarm can be silenced, but the visual alarm will remain on until the valve is reopened. This is a good check against a valve being left in the closed position. If a valve does not have a tamper switch, a method of notifying a responsible individual (other than the installing crew) that the valve is closed must be specified and carried out. One method is to have a tag in a guard station or some other constantly attended location that must be removed when the work shift ends. The point is that it is imperative that a method that satisfies the particular situation be devised.

Change of Occupancy

The obvious problem presented by a change of occupancy is an existing sprinkler system that was designed and installed for much less demanding conditions than are now required for the new occupancy. For example, the new system may involve high-piled storage of combustibles, flammable liquids, or machinery using flammable hydraulic fluid under high pressure.

The problem can usually be solved by "beefing-up" the existing system, but this may necessitate experienced engineering solutions. A system designed and hydraulically calculated to discharge a specified gpm per square foot over a specified area of operation may not be required with the change of occupancy to discharge double the gpm per square foot, and/or greatly increased area of operation.

A hydraulic analysis of the existing system may indicate a change of heads from 1/2" to 17/32", large orifice heads, changing some line size, and installing a new supply main to supplement the existing supply mains.

When it becomes necessary to retrofit an existing system to increase the design criteria of the existing system, it is not a project for the fire protection layperson. Only an experienced fire protection engineer who is well versed in hydraulic calculations and system design should attempt it.

Beefing up a pipe schedule system, or even converting a pipe schedule system to a hydraulically calculated system that provides a mandated design criteria, is doubly difficult, especially if the design criteria require a heavy discharge rate. Unless the water supply is exceptionally strong, it is almost impossible to hydraulically calculate a pipe schedule system to deliver a required density and area of application without changing some pipe sizes.

If a new addition to a sprinklered building is to obtain supply from the existing water supply, and the connection for the new system (which will have its own alarm valve and be a separate system) will be at the point where the underground enters the building to supply the existing system, remember that when taking two systems off the same supply, each must have its own control valve. The supply for the new system must be taken below the control valve of the existing system, and the new system will have its own dedicated control valve below the new alarm valve.

An architectural/engineering firm that obtains a project where the occupancy change necessitates upgrading the existing sprinkler system is strongly advised to retain an experienced fire protection engineer to handle this facet of the project.

Converting a Dry System to a Wet System

Some renovation projects involve a building that was originally not heated. In these cases, the sprinkler system was probably a dry system, but the renovation of the building is to include the addition of heat. The insurance carrier may recommend that adding heat to the building is to the owner's advantage, as he/she can now convert the dry pipe system to a wet pipe system, thereby eliminating the maintenance required on the air compressor, and the periodic inspections to ensure that the air pressure was being maintained on the system. In addition, the new storage occupancy in the warehouse would be more efficiently protected with the immediate discharge from a wet pipe system when a head fused. To convert a dry system to a wet system presents some problems, and naturally some cost, but the problems are not extremely difficult, and the cost is not excessive. Obviously, the most efficient way to convert the system is to remove the dry pipe valve, check valve, and connection of the fire department pumper connection from the riser. In other words "gut" the riser from the flange at the floor where the riser connects to the underground a few inches above the floor to the flange on the top of the dry pipe valve.

Starting at the flange on the underground located a few inches above the floor, build a wet pipe system riser with a control valve under the new alarm valve, if there was an inside control valve under the dry pipe valve, and a new alarm valve. The line (in all probability 4") to the fire department connection can be repiped to a connection to the riser above the new alarm valve, and the wet pipe main drain from the alarm valve reconnected to the discharge line of the drain from the removed dry pipe valve. If there was a water motor alarm, the piping to the existing water motor alarm can be reconnected to a retard chamber on the new alarm valve.

Auxiliary drains for low points on the former dry pipe system should be changed over to auxiliary wet pipe drains. This operation is a recommendation, not a mandate in the conversion.

The important part of the conversion is the hydrostatic testing of the wet pipe system. Sometimes fittings that have been dry for several years will tend to leak when under water pressure, but sprinkler fitters have a saying when a wet pipe system screwed fitting has a minute leak at a thread, "give it a little time and it will rust-in and seal itself." Seriously, the hydrostatic test of a converted system is extremely important; this is especially true if the system is quite a few years old.

Do not forget to state in the specifications and/or the contract that any leaks or damage to the system piping, other than the converted riser, will be corrected only after the owner authorizes the work based on an estimate of the cost. If the system is quite old, it is also recommended that some form of "hold harmless" clause be agreed on in writing to cover the contractor should a fitting blow out or break during the hydrostatic test and cause severe water damage.

Converting a Wet System to a Dry System

Converting a wet system to a dry system follows the same scenario as just described for a dry to wet conversion in terms of gutting the wet pipe riser and installing a dry pipe system riser. In this case, however, it is necessary to install a check valve downstream of the dry pipe valve to prevent the water that is pumped in through the fire department pumper connection from passing back through the underground piping. The existing fire department pumper line is repiped to a connection on the riser above the check valve and under the dry pipe valve. If there was an inside control valve on the wet pipe riser, a control valve is installed in the new dry pipe riser between the check valve and the connection of the fire department pumper to the riser.

Naturally, a compressed air supply must be provided from shop air or a dedicated air compressor. If the shop air is reliable and never shut down when the facility is shut down, an air maintenance device is installed in the air line to maintain a constant air pressure on the dry system.

The volume of the existing wet pipe system, soon to be a dry pipe system, must be calculated to determine if one dry pipe valve is adequate to handle the system, and/or if the volume is great enough to require some type of quick opening device to accelerate the exhaust of the system air. In addition, if shop air is found to be an unreliable air supply, the volume of the system will determine the compressor capacity. All auxiliary low point drains will have to be converted to approved dry pipe system drains. *Important consideration*: Wet pipe systems do not require the piping to be pitched to drain back to the main drain. The change in pipe size accelerates the drainage, and the only time the system is drained is to perform maintenance on the system, or when alterations are required on the system.

Dry pipe system piping must be pitched to ensure positive drainage of the water from the piping. Any residual water left in the piping will freeze, and could cause serious damage. It may be necessary, or it may be required by the insurance interest having jurisdiction, to pitch the lines, cross mains, and bulk piping on the wet system being converted to a dry pipe system. From a labor standpoint, this can be a rather expensive operation, and may even entail removal and replacement of riser nipples, rises, and drops on the system piping, and if the existing hangers do not have any adjustment, the replacement of hangers as well.

A thorough survey must be conducted of a wet system that is to be converted to a dry system, and the insurance carrier should be consulted to review the survey to avoid unexpected costs. Converting a wet pipe system to a dry pipe system is not as simple as converting the riser from wet to dry, and if at all possible the project should be contracted on a time and material basis, or at least a cost plus, not-to-exceed arrangement. This contract recommendation is mentioned only to convey the fact that the unexpected ramifications of this system conversion can have a serious cost impact if not given the cost estimating respect it deserves.

Hydrostatic testing of the completed system is, of course, very important. A pinhole leak that may "rust-in" on a wet pipe system can leak compressed air on a continuous basis, and cause the compressor to run continuously, or cause system trips. Again, be sure to include a clause in the specification and/or contract regarding the correction of leaks or damage to the existing piping during the hydrostatic test.

Trade-Offs

Trade-offs were mentioned briefly in Chapter 10, "Regulatory Agencies and Authorities Having Jurisdiction." A trade-off is the allowance of construction features in a building in exchange for a sprinkler system. Building codes have for years allowed larger floor areas and story height if the building was protected with automatic sprinklers. Some building codes allow unlimited floor areas if the building is sprinklered.

Some possible trade-offs are include greater travel distance to exits, lighter roof framing, and fewer fire walls, to mention a few of the many trade-offs being offered in modern construction in exchange for sprinkler protection. However, caution must be used in taking advantage of construction and life safety modifications in exchange for sprinkler protection.

Some codes offer a trade-off in one section of the code and then, in another paragraph, disallow the trade-off if a sprinkler system is required in a specific type of occupancy or construction. Rated corridor walls are one example in one particular code. For example:

The walls in exit corridors are mandated by this particular code to have one-hour fire resistance, but if a sprinkler system is installed, the walls are not required to have any fire resistance rating. The owner of this particular building decided to install a sprinkler system to take advantage of the fire insurance premium saving if the building was sprinklered, and the architect informed him that there would be a saving in not having to rate the corridor walls.

Another section of this code mandated that because of the occupancy classification of the building, sprinkler protection was required. The section relating to fire resistance of corridor walls states in an "exception" that if sprinklers are required by any other section of the code, the corridor wall trade-off is not applicable. In this case, the corridor walls had to be redesigned and rebid because the code requirements had not been reviewed thoroughly enough.

Building code trade-offs are sometimes necessary. For example, an architect cannot build a one-story building with the occupancy classification of a factory using material that is neither flammable nor combustible, and constructed of structural steel with no fire resistance, such as spray on fire resistive material, over 12,000 square feet in size. If, however, the building is sprinklered (and surrounded on all sides by yards of 60' or greater width (for fire department access), the area of the one-story factory is virtually unlimited. The unlimited area under the same conditions holds true for huge discount stores and grocery stores — a trade-off allowed by the building code in exchange for a sprinkler system.

It is obvious that the cost of a sprinkler system is offset by the construction savings in trade-offs, and as mentioned above, trade-offs are sometimes necessary in order to build a facility with the required square footage or increase in story height. *A note of caution*: Use sprinkler trade-offs of a sprinkler system for construction and life safety features very carefully. Sprinkler trade-offs for life safety construction features must be used with extreme caution and with sound professional judgment.

Remodeling Sprinkler System Notes

Questions often arise regarding sprinkler spacing on system remodeling projects. The following is a very brief description of some of the spacing and coverage guidelines.

Spacing

With new or remodeled sprinkler systems, the maximum distance between sprinkler heads or sprinkler lines is 15'. The minimum distance between sprinklers is 6'. By discharging water on adjacent heads, all heads are kept so cool that they would never obtain enough heat from the combustion to fuse. This is commonly referred to as "cold soldering."

It is easy to understand that the distance between sprinklers on a sprinkler line, multiplied by the distance between the sprinkler lines, will produce the square footage of each sprinkler head. If the design criteria states that sprinkler coverage will not exceed 100 square feet, and the sprinkler heads on a line are 10' apart, the sprinkler lines cannot be more than 10' apart. Simple to calculate, but this method must always be used once the maximum square footage of each head is determined by specification or fire code for the particular occupancy.

Remember, the heads can be spaced a maximum of 15', so if the head spacing is 130 square feet, and heads are being placed in a long corridor 8'-6" wide, they can be spaced 15' on center to arrive at 127.5 square feet.

If the corridor is 15' wide and the engineer is able to run the sprinkler line down the center, the heads can only be spaced at a maximum of 8'-6" to obtain 130 square feet or less.

The following issue seems to cause a great deal of confusion among those who do not deal with sprinkler design on a daily basis: If you have several sprinkler lines 12' apart with heads 10' on center to arrive at the required 120 square feet per head, consider that each head on each line must cover 6' between lines, and the head on the next line also covers 6', completely filling the 12' area between the lines.

Now consider that in the remodeling, a wall is placed on center between the sprinkler lines. The heads are 6' off the wall, so they are still covering the same area. This distance off the wall, or the area covered by the heads, must be considered as the 12' between lines, because the heads are covering the same 6'. For example:

> If a room is 10' wide, but there is a light in the center of the room, the sprinkler line cannot be run down the center line of the room, but must be placed 7' off the wall. The required head spacing is 130 square feet, and the line 7' off the wall is in reality 14' between imaginary sprinkler lines. When 130 is divided by 14, the answer—9.2'—is the maximum space one can have between heads on the line in the room in order to be under the 130 square feet per head limit.

> One more example: a room 10' x 12'. Head spacing required 130 square feet. Light and diffuser in the center of the room. If a head could be placed in the center of the room, the coverage would be 120 square feet, but the only place to put a sprinkler is 5' off each side wall, but 7' off the end wall, which places it 5' off the other end wall.

> What we have in theory, is heads 14' apart, and lines 10' apart. In other words, heads 10' apart, and lines 14' apart. No matter how you look at it, the head coverage in this room is 140 square feet. Obviously this exceeds the maximum allowed, and it will be necessary to put two heads in the room.

When spacing heads, the rule is that the distance of the end head on the line must not be more than half the distance between heads on the line off the wall. If the heads are 10' apart on the line, the end head cannot be more than 5' off the wall—or beam if the beam should form an obstruction. Now this end head can be less than one half the distance between heads on the line, but never more than half the distance between heads on the line.

The same is true for sprinkler lines. The lines cannot be more than half the distance off the wall (or beam) than the distance between the lines. Lines 12' on center cannot be more than 6' off the wall (or beam). They can be less than 6', but never more than this distance.

Remember that heads cannot be more than 15' apart, maximum. Therefore, the end head can never be more than 7'-6" off the wall (or beam), nor can the lines be more than 7'-6" off the wall (or beam).

This explanation may make the spacing in the rooms a little clearer. Off the wall distance is in reality half the distance between heads or lines, and the square foot spacing is based on this concept. In the rooms used in the examples, we doubled the distance of the head (or heads) off the walls, and multiplied this by double the distance of the head off the other wall to get the square footage.

> *One last room example*: A room 12' x 12'. Maximum head coverage required 120 square feet. There are no ceiling obstructions, so the heads will be placed in the center of the room, 6' off each wall.

This 6′ dimension defines that the "lines" are 12′ apart, so we know that we cannot go over 10′ between heads. We could go 10′ between the heads, but that would put the two end heads 1′ off the end walls. To give a little better room coverage with the end heads by backing them a little further off the end walls, one could space 8′ between the two heads and have 2′ off the end walls.

The distance between these two heads can be spaced at other intervals as well, so long as the two heads are not more than 5′ off the end walls, so the 120 square feet per head standard is maintained. This is an extreme statement used to make a point, because the heads should not be placed any closer than 6′ apart, and this cannot be done because they would then be more than 5′ off the end walls.

Another scenario: If there are obstructions on the ceiling, one could space one head 5′ off one end wall, and this would put the other head 1′ off the other end wall. We still have not exceeded the square footage requirement because the one head is only 5′ off the end wall.

When discussing these room spacings, the wall was followed by a "beam." If the beam prevents the head discharge from being directed half the distance to the next head, it can be treated as a wall in terms of its obstruction.

In a building with 24′ bays formed by deep, solid beams, these beams can be considered walls when spacing sprinkler lines. In the example building, it would be best, if possible, to space the lines 12′ on center, with each line 6′ off the beams in the bay. Now the head spacing required is divided by 12′ to arrive at the maximum distance between heads. With 12′ between lines and 12′ head spacing, a maximum of 10′ is allowed between heads, with the end heads not more than 5′ off the walls of the building.

Unfortunately, one cannot always get spacing as "neat and tidy" as 10′ × 12′. The 12′ spacing in this example should be fixed, because it is always good engineering practice to "quarter point" the lines—space the lines half the bay distance between the lines, and one quarter the bay dimension for the lines off the beams. Quarter pointing can be used in rooms to get a more symmetrical head layout if, and only if, one does not exceed the required square footage of each head.

It is hoped that this explanation will offer some insight into the spacing of sprinkler heads and lines to accomplish coverage of the required square footage. To the sprinkler designer or fire protection engineer it becomes second nature, but to the layperson, it is very difficult to realize that when you have only one head in a small room, you are really looking at a head on a line half the distance from the next head that really doesn't exist.

Practice is the only solution, and practice before it becomes a real situation. Revamping and remodeling sprinkler systems where new partitions are installed present very real and common situations.

CHAPTER TWELVE

FIRE PROTECTION SYSTEM SPECIFICATIONS

CHAPTER TWELVE

FIRE PROTECTION SYSTEM SPECIFICATIONS

Fire protection specifications are extremely important documents—from the standpoint of both project performance and quality assurance. The fire protection specification tells the contractor exactly what the designer wants in the way of equipment, material, installation, and overall project quality.

Fire protection contractors may sometimes question why a certain product was specified, when an alternative would reduce costs. Fire protection contractors may also criticize specifications with the attitude that the designer is not so knowledgeable about specifying a fire protection system. In fact, many fire protection contractors do not realize or understand the engineer's role in preparing the specification, and have the attitude that the specification was written at the last minute based on the first time the engineer had a chance to look at the project. These contractors do not realize the hours spent at meetings, conducting project research, coordinating the work with the other disciplines, and making calculations and evaluations—all of which went into the reasons for specifying exactly what is in the specification.

On the other side of the coin, most fire protection contractors are professionals in what they do, and are, of course, in business to make a profit. The engineer who writes the specification does not always realize that he or she prepares the specification in an entirely different environment from that of the contractor.

When the fire protection contractor receives the project specification, he must take this document written by a professional stranger with unknown fire protection expertise, and convert it into a bid that will obtain the contract for his company, while complying with the specification, and with the codes and standards that he knows and works with on a day to day basis.

This is not an easy task in today's competitive marketplace: To come up with a low bid and still give the client a quality installation. It becomes very evident to the professional fire contractor when he reads the specification whether or not the engineer has done a thorough job of investigating all aspects of the project. If there are concerns, he may feel compelled to add in a cost allowance for extras.

Engineers who prepare fire protection specifications would benefit if they spend a minimum of six months as a specification apprentice working with a fire protection contractor. This is one way to discover the cost impacts of potential plan revisions.

This is not a criticism of the engineers who prepare fire protection specifications or the fire protection contractor. The point is that each side should understand the other's concerns and objectives. The engineer who writes the specification should realize how the fire protection contractor uses it, and vice versa.

The design engineer's specification is more than an exercise in demonstrating his knowledge of fire protection systems; It is a real-life document that should both provide the client with a quality project, and give the installing contractor the opportunity to be successful.

One thing for an engineer to remember: *You cannot fool a professional fire protection contractor.* These professionals know by reading the first paragraph of a specification whether or not the preparer knows fire protection systems, and understands what fire protection contracting is all about. The fire protection specification writer, cannot "con" an experienced fire protection contractor with pretended expertise in the specification or with "cover all" statements.

Designers should always keep in mind when preparing the specification that someone is going to read it who has a great deal more "hands-on" knowledge of fire protection systems, regardless of whether you have a "P.E." after your name or other prestigious college degree hanging on the wall. This thought should keep fire protection specification writers honest, humble, and precise.

Specification Comments

References
Under "References," be sure to give the contractor the specific information he or she needs. Is the client insured by an entity other than Factory Mutual? If so, list Underwriter's Laboratories and the National Fire Protection Association, and if you know for certain that Industrial Risk Insurers is the insurer, list the IRI.

The statement, "authorities having jurisdiction," does not tell the contractor what specific requirements will apply.

If the client is insured by Factory Mutual, so state in the specification, but don't add NFPA. The Factory Mutual Data Sheets that are equivalent to NFPA Fire Codes will provide the necessary requirements, and the FM Data Sheets do not always agree with NFPA Fire Codes in all areas. FM has some specific requirements and exceptions in their Data Sheets even when they use an NFPA Fire Code word for word.

Applicable Standards
Here again, many specifications will list the gamut of standards that are to be applied to the project. These usually include Underwriter's Laboratories, Factory Mutual, and the National Fire Protection Association.

If this is a Factory Mutual insured client, list Factory Mutual Data Sheets and the Factory Mutual Approval Guide. The Factory Mutual Guide is equivalent to the publication that lists UL equipment, but it indicates only the products and systems that have the Factory Mutual approval.

There are some products, systems, equipment, and devices that are FM Approved, but not UL listed. By indicating both of these approval bodies, the specification is allowing the contractor to use any item that is FM approved or UL listed on a Factory Mutual project.

It is true that Factory Mutual will accept some items that FM does not include in their Approval Guide but are UL listed, but when this discrepancy becomes apparent on a project, it should be handled item-by-item with approval from the designing engineer and Factory Mutual. Make this list of Applicable Standards specific to the project.

Design Criteria

The quickest way to have a contractor lose confidence in the professionalism of the specification is to have the specification state hydraulic design criteria (gpm per square foot and area of operation), and then in another section state that the piping shall be sized by the pipe schedule. The project must be either hydraulically calculated or sized by the pipe schedule method.

If the specification states that the system is to be hydraulically calculated, the design criteria should also be stated, and should include the flow data the installer is to use. Without both of these items the contractor cannot design the system or size the piping.

If the engineer has prepared a preliminary layout and sized the piping on this preliminary layout by hydraulic calculations, the specification should not "lock" the contractor into using this layout or the pipe sizing per se.

Sometimes it is necessary for the engineer to prepare a preliminary layout and to size the piping by hydraulic calculations so that the estimator can have a base to work from, or so that the structural engineers have an idea of piping weight. Nevertheless, the contractor should be given the option to redesign and recalculate the system subject to the approval of the engineer and in strict compliance with the specifications.

This flexibility may result in a more efficient and, for the client, *less expensive system*, because every contractor will be making an effort to comply with the specification, while providing a system that will make the bid competitive. As quoted previously, the old saying applies as well: "Give a plan to five sprinkler contractors, and you will get five different approved layouts."

Should all control valves be supervised? If so, make this statement, but follow it by specifying how the control valves are to be supervised. NFPA provides several means for the supervision of control valves, and Factory Mutual and IRI usually have very specific requirements. Some specifications include the statement that the requirements of the public health authority having jurisdiction shall be determined and complied with regarding public water contamination. The fact is that the engineer must determine what is required or not required. When it is determined that backflow prevention is required, state in the specification which type of backflow preventer is required. This should not be left up to the contractor, especially during the bidding period, and it is not a requirement to be determined by the estimator.

Some specifications use the statement: "Cathodic protection is required." The engineer is the only qualified person to determine this requirement. If it is required, so specify; if not, so specify. The responsibility is in the hands of the specification writer.

Metering is another item that is the engineer's responsibility to determine. If metering is not required, so specify, but if it is required (based on investigation) specify which type of metering is required.

Some specifications will include a section covering design of the hose stations. If the engineer has determined that hose stations are required, he should specify exactly what equipment shall be used, and how the hose stations are to be supplied. Are they to be small hose supplied from wet pipe sprinkler systems, or hose stations supplied from a standpipe that is also supplying sprinkler systems? Hose type and quality, type of nozzle, rack or reel, and hanging arrangements should all be stated. There are times when hose stations are not required. If this is the case, do not plug in a section in the specification covering hose stations "if required" state that no hose stations are required.

When the system is a dry pipe system, do not specify a quick opening device "if required." The engineer should know approximately the volume of the dry pipe system, and whether or not the exhauster or quick opening device is required. The engineer has to know the approximate volume of the dry system in order to determine if one system will accommodate the project.

Always specify that the shop drawings, or installation plans, shall indicate the dry pipe system capacity. This is very important when the shop drawings are submitted for approval.

Hydraulic Calculation Specifications

As previously mentioned, design criteria must be determined by the engineer and clearly stated in the specification. If several areas of the project facility will have different design criteria requirements, they can be indicated on the plans, and the specification can read: "Design criteria relative to water density and area of operation shall be as indicated on the plans."

An important point, often left out of a specification for a dry pipe system, is the 30% increase in area of operation that may be required when hydraulically calculating a dry pipe system. If the specification reads simply "so many gpm per square foot over so many square feet of operation," the contractor is not sure if the 30% is included or if he is to add it. This will make a big difference on his estimate sheet.

Velocity pressure is sometimes excluded by the specification, stating that velocity pressure shall not be used in the hydraulic calculations. Whether or not to allow the contractor to use velocity pressure in his calculations is sometimes a judgment call by the engineer, and sometimes a requirement of the insurance interest having jurisdiction. It is a good idea to verify the insurance carrier's requirements on this issue before it is written into or left out of the specification.

If velocity pressure is strictly a judgment call by the engineer, that judgment will have to be based on the project conditions, and whether or not this "safety factor" can be eliminated in the calculations. Water supply is a big factor and some water supplies, coupled with the specified system, will not be capable of satisfying the design requirements without using velocity pressure.

This discussion on velocity pressure also applies to velocity of the water in the piping. The velocity (feet per second) of water in the piping can also be left to the judgment of the engineer, or can be a specific requirement of the insurance carrier on the project. If the specification is going to restrict the calculations to a specific velocity, be absolutely sure that the velocity restriction so specified will meet the approval of the insurance carrier, and that the specified velocity has not defeated the calculation results relative to supply vs. demand.

Do not specify that velocity shall be in accordance with NFPA requirements, because NFPA does not provide requirements relative to velocity. Nor shall the specification writer state that velocity will comply with the insurance interest having jurisdiction. The engineer/spec writer must make this professional determination and so specify.

It is always good practice to include in the specification some wording that the installing contractor shall not add heads or institute pipe changes in direction or size without submitting revised calculations. This is a statement that will prevent the installing crew from changing piping layout in the field to avoid an obstruction, without realizing that these changes can affect the hydraulic calculations.

Acceptable Manufacturers

Specifications usually include a statement referring to all those manufacturers listed in the Underwriter's Laboratories publications, and all those listed in the Factory Mutual Approval Guide. Some manufacturers' products that are listed in the UL publication are not listed in the FM Approval Guide. One example is a commonly used valve in fire protection systems. It is UL listed, but is not FM approved. If the project is FM insured, this specification gives the contractor the option to use either valve, which may lead to a denial of FM approval of the system. If the project is FM insured, use the Factory Mutual Approval Guide. If the project is not an FM risk, use the Underwriter's Laboratory publications. As mentioned before, on an FM project the contractor can usually get FM to approve the use of a UL listed product if for some reason the FM approved product is not available, but the contractor should not be given an open choice on an FM project.

Equipment and Material

Here again, specifications usually include the statement that all equipment, devices, and accessories shall be either UL listed or FM approved. Determine which is applicable and so specify. It is good practice to limit the contractor to the use of products from a single manufacturer when two or more products of the same type are required by the system or systems. Do not allow the contractor to "mix and match" with products he might have on the shelf. This does not produce a quality installation, and when replacement parts or units are required, it can cause additional problems.

Most specifications will leave the galvanizing or corrosion protection of piping up to the contractor by stating that the piping shall be protected in areas that require it. It would be a much better and more professional practice for the engineer to determine where these areas exist, and then either indicate them on the plan or identify them in the specification. This is really the responsibility of the engineer who should be quite familiar with the entire project and project occupancy. The desired trim and accessories should be specified for valves, alarm valves, dry pipe valves, and deluge valves. This applies not only to the standard trim that comes with the complete valve, but anything that might be considered extra to the standard trim. For example, a retard chamber, a pressure switch, a pressure switch with two sets of contacts, a water motor alarm, an electric alarm, etc. This is good practice, because it requires the estimator to include all these items in his estimate.

If a dry pipe valve house is required, it should not be included in the fire protection contractor's specification unless it is intended that he include the construction in his bid. Usually this house will be designed by an architect and should be specified in the general construction specification. This is mentioned because some specifications will state that the dry pipe valve shall be installed in a dry pipe valve house, and the fire protection contractor is not sure whether he should include this in his bid, or at least the equipment that must be installed in the valve house, such as the thermostatically controlled heater and low temperature alarm.

If the system design calls for a *vane-type water flow switch*, be absolutely sure to specify that the flow switch shall be the instantly recycling type, so that flows that do not activate the device (because they occur within the retard time frame) are not cumulative. The water flow switch should have a field adjustable retard, but it is the responsibility of the engineer preparing the specification to determine to the best of his ability the retard time. Make sure by specification and/or plan detail that the required pressure gauge and main drain are included in the water flow detector riser.

Sprinkler Heads

The usual statement in specifications that all sprinklers shall be UL listed or FM approved, a blanket statement, is not so problematic when referencing sprinkler heads used on "standard" sprinkler systems, but problems do arise when the system involves a newly developed head that may have a UL listing, but has not yet been approved by Factory Mutual. Also, in some systems, foam systems for example, sprinkler heads have FM approval, but no UL listing. To be more professional, and subject to fewer conflicts, the engineer writing the specification should be sure the heads on his or her project carry both the UL listing and FM approval.

If there are suspended ceilings with concealed piping and pendent heads, and the heads are to be a special type, it may be necessary to specify a manufacturers' model number "or engineer approved equal." This is because a pendent head specified as a "flush head" does not necessarily mean that the specifier will get the type of flush head expected. There are several types and patterns of flush heads, and several that are referred to as "flush heads" that are really not what one might expect.

The decision to require corrosion-proof heads should be the responsibility of the engineer. He or she has lived with the project since its conception, and should know where these corrosive problem areas exist. The contractor should then be informed by specification and/or plan notations. When the shop drawings are received for review, or when the installation is inspected for final approval and certain areas are recommended to have corrosion-proof heads, a specification item will eliminate any discussion. If the "as required" was the specification item, the engineer could be faced with a difference of opinion with the contractor.

Make sure the specification for spare heads includes the requirement that spare heads of every type used on the project be included in the spare head cabinet, and that a sprinkler wrench specifically adapted to removal and replacement of every type of head used on the project be included in the spare head cabinet.

Temperature rating of sprinkler heads can be a tough call in a specification. The engineer should know which areas will require high temperature heads, and should specify these areas or indicate them on the plans. One situation that does not require a judgment call is when the insurance carrier mandates the temperature rating of the heads; this should definitely be specified. In new buildings, the engineer should have a fairly good idea of where unusual heat conditions will exist, or where heat producing equipment will be located. This information should be included in the specification or so noted on the plan. Unless the insurance carrier requires a specific temperature rating, NFPA and FM offer excellent guidelines relative to head versus ambient temperature conditions.

If there is no way to determine the maximum temperature of an area, it is a good idea to have the contractor furnish and install a recording thermometer. The resulting temperature readings will determine the proper head temperature. This may require a return trip to the job by the contractor, with ladders and a crew if the installation is complete before the temperature determination is made. This extra expense should be discussed and agreed upon by all parties concerned, since it could have been avoided if the engineer had determined the heat conditions and head temperature rating when the specification was prepared during design.

Be sure that the specification for the type of heads does not conflict with NFPA, FM, or the insurance interest having jurisdiction. For example:

> A rack sprinkler project was specified according to the direction, desires, and mandates use of the client's fire protection personnel. The specification read that the system was to comply with NFPA requirements in NFPA Standard 231C, the rack protection standard. The heads required by the client's personnel did not coincide with the requirements of the type of heads mandated by 231C. As a result, the contractor informed the engineer that he would not stamp the plans because they were not in accordance with NFPA requirements, whereas the specification stated that the plans had to meet NFPA requirements.

This example is an extreme, but actual, case which points out that the specification should never conflict with the requirements of the code or standard referenced for any specific item.

Drains and test connections are not a major specification problem, but if the specification states that the drain shall discharge to the closest drain outlet, and the engineer knows that the closest drain outlet is a floor drain, be sure to specify that the drain line be carried to within, say 6" of the floor drain. Otherwise, some drain discharge lines will spill out on the floor far from the floor drain, and result in water damage, or at least an unnecessary mess. The sprinkler engineer should always inform the plumbing engineer of this drain discharge and the amount of water anticipated, so the floor drain can accept the volume without causing an overflow.

Drains tend to be bypassed on plans and in specifications. There is a sort of a "leave-it-up-to-the-contractor" attitude. This is definitely wrong because unless the contractor is extremely conscientious, he will be inclined to arrange the drain discharge in the least expensive manner. Specify exactly where the drain should be located, and where it should discharge.

Under the category of control valves, there are one or two cautions. When specifying wall post indicator valves, be sure to specify the specific type. There are *recessed, fully recessed,* and *totally exposed* wall post indicators. Depending on which type is desired, the engineer must coordinate the design and specification provisions with the architect, and possibly the structural engineer, depending on the type of wall structure involved.

When a wall post indicator is to be used on a project, always specify, or better yet detail on the plan, the location of the flange and spigot piece off the wall. Each type of wall post indicator requires a certain distance off the wall to facilitate installation.

And speaking of the flange and spigot piece, specify that the flange shall be located 6" to 8" above the finished floor. Having the flange of the flange-and-spigot piece a foot or so above the finished floor is not acceptable.

All control valves on the site of a facility should open in the same direction. If the installation is an addition to a sprinklered facility, determine which way the existing valves open, and specify that the new valves open in the same direction. This is extremely important for outside control valves installed on the underground. If a valve is turned in the wrong direction to open, it can be seriously damaged, and may become inoperable.

Pipe supports and attachments must be UL listed or FM approved. Approved "standard" hanger requirements are adequately covered by NFPA and FM publications. There must not be any conflict in the specification between a listed or approved fire protection pipe support and a hanger designed by an engineer who has used his hanger arrangement on process piping. The hangers for fire protection must be UL listed or FM approved, including rod size, or the system will not comply with NFPA or FM requirements.

The engineer should recommend in the specification that no sprinkler piping be supported from the bottom chord of bar joists, because too much weight on the bottom chord of the bar joist can turn the bar joist out of the vertical position. Always include an earthquake zone hanger requirement, and when a fire department connection is specified, always include the statement that the hose threads shall comply with the specifications of the local fire department. Or obtain from the local fire department their hose thread specifications and include those in the specification. *Know your project thoroughly before you specify the fire department connection,* and then specify exactly what is needed on the specific project. Always specify that the ball drip drain valve be installed to maintain the fire department connection in a dry state. Be aware that this automatic ball drip drain will, at times, discharge a small amount of water and provide for this occurrence.

Testing

Specify that all tests shall be conducted by the installing contractor, and specify who shall provide all labor and equipment to conduct these tests. Always specify that the contractor shall notify all concerned parties (owner, engineer, insurance carrier, fire marshal) five days or more in advance that the tests will be conducted. If the situation requires more advance notice, inform all concerned parties accordingly so that they may witness the test. This notice should also be issued several days before the flushing of the underground, so that this operation can also be witnessed by all concerned parties.

Specify that the use of waterglass shall not be permitted. Waterglass is a liquid that can be injected into the system piping when there are undetected leaks. When the waterglass reaches the pinhole leak, it leaks out, and when exposed to air, hardens and seals the leak. Waterglass can cause some major problems to system equipment and devices, and its use should never be permitted. Specify that following the testing of a system, the system shall be returned to a functional and operational condition at no extra cost to the owner.

Any retesting that is required because of failure of any test shall be specified to be conducted at no additional cost to the owner, and any corrections or repairs to the system to permit retesting shall also be performed at no cost to the owner.

Unless the specification is complete, specific, and "tight," the engineer will not have "a leg to stand on" if there is a dispute with the contractor and/or client when the shop or installation plans are reviewed for engineer approval, or when the installation is complete. Field supervisory personnel need a good, well prepared, specification in order to make sure the contractor is complying with the specifications.

A specification that merely states that the system shall be installed in accordance with NFPA could be in for a lot of problems and cost impacts. The problem is not that NFPA is not an effective guide, but rather that there are many areas where a contractor can take advantage of a less expensive arrangement that meets NFPA requirements, but definitely does not meet the client's desires.

Summary

The fire protection specifier must be made aware of the serious problems that can be caused by blanket statements in a specification. This individual must not only thoroughly understand, but also convey, an understanding of the project requirements and authorities having jurisdiction, and of exactly what is required on the project.

GLOSSARY AND ABBREVIATIONS

GLOSSARY

across the line electric motor starting
The electric drive motor of a fire pump that starts at full voltage up to speed almost instantly.

air maintenance device
A device installed in the line from shop air to control the air pressure supplied to a dry pipe system or a supervised pre-action system.

aqueous film forming foam (AFFF)
A quick knock-down foam that spreads an aqueous film over the surface of the flammable or combustible liquid.

branch lines
The pipe that supplies the sprinkler heads.

bulk main (feed main)
A supply pipe without any connections. For example, pipe from the alarm valve or dry pipe valve to connection of the sprinkler system.

butterfly valve
A very thin control valve. Some butterfly valves are not both UL listed and FM approved.

cold soldering
When heads are installed so that one head discharge will cover another adjacent head, the discharge will keep the adjacent head cool and prevent it from fusing. Sprinkler heads cannot be placed closer than 6' in order to prevent cold soldering.

combustible liquid
A liquid that has a flash point at or above 100°F. As with flammable liquids, combustible liquids are broken down into three classifications depending on their flash points.

continuous flow hose
Hose that is stored on a reel and can be operated no matter how much hose is pulled off the reel. With regular folded hose, it is necessary to pull the entire length of hose off the rack before it can be used.

corroproof heads
Sprinkler heads or nozzles that are coated with a wax material or lead to prevent corrosion when the heads or nozzles are installed in a corrosive atmosphere or in an area where they are exposed to weather.

coupling guard
The cover over the coupling on the shaft between the driver and the fire pump to prevent personnel injury should someone not realize that the shaft is rotating at high speed.

cross main
The pipe that supplies the sprinkler lines.

deflector
The water discharging from the orifice of a sprinkler head impacts on the deflector to form the discharge pattern of the head. The deflector of an SSU is large, that of an SSP is small.

down
Indicates (on a sprinkler plan) a pipe going from one story of the building to the next lower story.

drop
On a sprinkler plan, "drop" indicates a pipe that runs from a higher elevation to a lower elevation of the system, within the same story.

dry powder
A dry powder agent that is used to suppress combustion involving combustible metals.

escutcheon
The ceiling plate that covers the hole in the suspended ceiling where the drop nipple penetrates the ceiling to supply the pendent sprinkler.

eutetic solder
The fusible element of the sprinkler head that is a compound of material designed to become soft, or melt, at a specific temperature, providing the temperature rating of the head. The most common temperature ratings are: $165°$, $212°$, and $286°$, but nearly every manufacturer supplies a vast range of heads with different temperature ratings.

extinguishers
A device used to suppress a fire manually. Extinguishers are classified according to the type of combustion they best extinguish:
> "A": Ordinary combustibles such as wood, paper, etc.
> "B": Flammable liquids.
> "C": The agent in the extinguisher can be safely applied to energized electrical wiring and equipment.
> "D": For use with combustible metals.

Factory Mutual (FM)
An independent testing organization. Any equipment, device, or system manufacturer who wishes to display the FM label submits samples to the organization for extensive testing.

feet per second (FPS)
The term used to describe the velocity of the water in a piping system.

fire watch
The appointment of an individual to remain in an area unprotected by the existing sprinkler system to watch for any fires. Used when it is necessary to shut down an existing sprinkler system to make repairs, remodel, or make a connection.

flammable liquid
A liquid having a flash point below 100°F. There are several definitions of a flammable liquid based on their flash points and boiling points (See NFPA Standard 30).

flash point
The minimum temperature at which a liquid will give off sufficient vapor concentration at the surface of the liquid to form an ignitable mixture with air.

flow control valve
A deluge valve that can be opened and closed automatically. Controls the discharge regardless of the supply pressure—within limits.

heat actuated device (HAD)
A pneumatic rate of rise detector. A hollow, almost football shaped, device. When the combustion heat rises at least 15°F per minute, the air in the device expands and activates an alarm device.

heating, ventilation, air-conditioning (HVAC)
Refers to the ductwork, equipment, fans, motors, units, etc. that provide the creature comfort for the building.

hot start
The device that maintains adequate temperature in the diesel drive of a fire pump when it is not operating so that the engine will start quickly when called upon to activate. It is normally an electric device attached to the engine.

hydraulic placard
The metal embossed or painted sign that must be installed on the system riser to indicate the hydraulic design criteria of the system.

hydraulic reference points
Indications on the plans—numbers or letters—that are identical to the numbers or letters on the hydraulic calculation sheets.

infra red detectors
Flame detectors that are activated by the infra red radiation of the flame.

inspector's test connection
A test connection to test the operation of all the system alarm devices. When the inspector opens the test connection valve, the water must flow within 60 seconds. If this flow does not occur in 60 seconds, it indicates that the dry pipe system will require some device to exhaust the air pressure, or a device to admit air to the intermediate chamber of the dry valve.

iron pipe thread (IPT)
The thread machined on the end of a piece of pipe.

jockey pump
A pump that maintains a constant pressure on system piping, and is sometimes referred to as a pressure maintenance pump.

local application
Applying a fluid agent directly on the hazard when the hazard is not enclosed in an area suited for total flooding.

machine thread rod (MTR)
A hanger rod that has been machined to have a thread for a rod coupling. Sometimes the entire length of the hanger rod will have a running machine thread which can then be cut to the desired length. Eliminates having to cut a plain end hanger rod and then machining a thread on the end or both ends.

monitor
A foam or a water monitor is a device that can deliver large quantities of water or foam on a fire. Monitors are permanently fixed devices, and can be mounted on a platform or on one outlet of a fire hydrant. May have either a fixed discharge position or an oscillating pattern.

national pipe thread
A standard that indicates the size, dimension, and number of threads per inch for a pipe thread. When ordering a fire department pumper connection, a siamese connection, the specification must always read: "siamese connection threads shall comply with local fire department specifications."

orifice
The discharge opening of a sprinkler head or nozzle. A standard sprinkler head has a 1/2" orifice, a large sprinkler head has a 17/32" orifice.

outside screw and yoke valve
A control valve which, when open, has the stem extending several inches above the wheel so its condition (open or shut) can be determined at a glance.

overload
This is the term that is used when a fire pump is discharging more water than it is designed to discharge.

post indicator valve
A post that extends a few feet above grade and is attached to a non-rising stem gate valve in the underground pipe. The post has a window that indicates whether it is "open" or "shut."

post indicator valve assembly
A post indicator attached to the gate valve installed in the underground. The post indicator valve assembly is a one piece unit, where the post indicator is two—a post indicator and a gate valve. The post of the post indicator valve assembly has the indicating window.

pre-mix foam solution
A solution of foam and water in a tank produced by introducing a measured amount of foam concentrate into a known quantity of water. Foam concentrate in a pure state has an almost indefinite shelf life (although it should be tested annually), but foam concentrate in a pre-mix solution has a limited shelf-life for maximum fire control efficiency.

ratio controller
A device that uses a venturi principle to inject foam concentrate into a water supply. The throat in the device increases the velocity of the water and thus lowers the pressure so that foam concentrate can be drawn into the water supply.

reducing coupling (RC)
A pipe fitting used to connect two pieces of pipe with different diameters or sizes.

riser
The pipe that rises vertically from the valve to the bulk main or system piping. The pipe rises from floor to floor to supply the sprinkler, or standpipe systems on each floor.

reduced voltage electric motor starting
When the power source to the electric drive motor of a fire pump cannot accommodate the tremendous surge of power to an across the line motor, a reduced voltage motor starts in stages so as not to place a massive drain on the electric power source.

riser nipple
If the cross main is at a lower elevation than branch lines, the riser nipple is the pipe from the cross main to the elevation of the branch line. If this piece of pipe is greater in length than 6″, it is referred to as a *riser pipe*.

sectional valve
A control valve that sections off an isolated zone of underground piping. This facilitates closing off a piping of the underground system for repairs while leaving the remainder of the underground supply operational.

sequence pump starting
A sequence starting device that allows a few seconds between the starting of motors.

skeleton lines
When all the branch lines are exactly the same length and size, one or two complete lines are shown, but the other lines are skeletonized by just showing the starter piece off of the cross main supplying the first head off the cross main.

smooth bore brass bushing
A 1″ × 1/2″ smooth bore brass bushing that is screwed into the 1″ discharge of the inspector's test connection to give the flow equivalent to one sprinkler operating.

splash shield
A metal shield between the fire pump and the electric motor driver to prevent water leaking from the fire pump from damaging the electric motor.

spray sprinkler pendent (SSP)
A sprinkler head used in the pendent position.

spray sprinkler upright (SSU)
A sprinkler head used in the upright position.

starter piece
The piece of pipe that "starts" the branch line. The piece of pipe from the cross main or riser nipple to the first head.

tamper switch
A switch on a system control valve that prevents tampering with the valve.

test blanks
When a portion of a system is to be tested without testing the entire piping system, a test blank is installed to isolate the area to be tested from the remainder of the system. These test blanks are identified by red painted lugs that protrude beyond the flange to clearly indicate their presence.

thrust blocks
A calculated mass of concrete poured against a change in underground pipe direction, hydrant, or valve to prevent the thrust of the water in the piping from separating the tee, ell, valve, or hydrant from the piping.

thread one end (TOE)
Indication that the sprinkler fitters will have to take the measurement in the field, cut the pipe to length, and thread the plain end.

top of steel (TOS)
When a dimension is indicated "TOS" it means just what it says—the dimension is the top elevation of the steel joists, beams, or truss.

total flooding
Filling an enclosed area with a specified concentration of a fluid agent.

ultraviolet detectors
Flame detectors that are activated by the ultraviolet radiation of the flame.

Underwriter's Laboratories (UL)
An independent testing organization. Any equipment, device, or system manufacturer who wishes to display the UL label indicating the product is "UL listed" submits samples to the UL for extensive testing.

velocity pressure
The pressure exerted by the velocity of water in piping.

wafer check valve
A very thin check valve.

water flow alarm device
This can be a pressure switch that is activated when water flows in the system, and water is applied to the switch to initiate an electrical contact. Can be a paddle-type water flow switch device that is activated when water flows in the system and deflects the paddle inserted in the pipe.

water motor alarm (WMA)
A mechanical bell that is operated off of system water pressure. When water flows in the system, a pipe carries water under pressure to an enclosed "paddle wheel" on the interior wall of the building, and rotates a shaft that passes through the wall to turn a striker against a bell mounted on the exterior wall.

wet tap
A connection to a water main that is under pressure without shutting down the main.

ABBREVIATIONS

AFFF or FFF	aqueous film forming foam
ASTM	American Society of Testing Material
FM	Factory Mutual
FPS	feet per second
gpm	gallons per minute
HAD	heat actuated device
HI	Hydraulic Institute
HVAC	heating, ventilating, air-conditioning
IPT	iron pipe thread
IR detector	infra red detector
IRI	Industrial Risk Insurers
ISO	Insurance Service Office
MTR	machine thread rod
NEMA	National Electric Manufacturers Association
NFPA	National Fire Protection Association
psi	pounds per square inch
psia	pounds per square inch absolute
R.N.	riser nipple
R.P.	riser pipe
RC	reducing coupling
SSP	spray sprinkler pendent
SSU	spray sprinkler upright
TOE	thread one end
TOS	top of steel
UL	Underwriter's Laboratories
UV detector	ultraviolet detector
WMA	water motor alarm

APPENDIX

Sprinkler Comparison

This brief table summarizes the characteristics, usage, "K" factor, reference sources, and restrictions, of basic and special application sprinkler heads. The reader is cautioned to obtain experienced fire protection engineering assistance, particularly for the special application heads (large drop and early suppression), which have specific restrictions and application requirements. The application of these heads requires a thorough understanding not only of the design criteria that must be applied, but also the limited use of these heads—to protect against only those hazards for which tests have proven their efficiency and effectiveness.

General Use Sprinklers

Sprinkler Type	Characteristics	Usage	"K" Factor	Reference	Restrictions
Upright Spray Sprinklers (SSU)	• Large deflector • Water directed up out of orifice against deflector. • 1/2" orifice, 1/2" NPT and 17/32" orifice, 3/4" NPT. • All temperature ranges, mfg., temp. and date of mfg. on head.	• General wet and dry sprinkler system usage on all systems used in upright position. • With fusible element removed, can be used on deluge (open head) systems.	1/2" orifice 5.2 to 5.8 17/32" orifice 8.0	NFPA FM UL	• Not to be used in the pendent position.
Pendent Spray Sprinklers (SSP)	• Small deflector. • Water directed downward from orifice against deflector. • 1/2" orifice, NPT and 17/32" orifice, 3/4" NPT. • All temperature ranges, mfg., temp. and date of mfg. on head.	• General wet pipe sprinkler system usage. • Also for exposed or concealed piping. • Standard and ornamental.	1/2" orifice 5.2 to 5.8 17/32" orifice 8.0 (average)	NFPA FM UL	• Cannot be used on dry pipe and pre-action systems. • Can only be used in the pendent position.
Sidewall Sprinklers	• Special deflector designed to discharge most of water away from nearby wall. • Upright and horizontal type.	• Primarily used in light hazard occupancies. • Installed at a wall to throw water out into room.	5.2 to 5.8 1/2" orifice 8.0 17/32" orifice	NFPA UL	• Spacing restrictions depend on width of room, distance down from ceiling and from wall to sidewall sprinkler. • UL Listing.
Extended Coverage Sidewall Sprinklers	• Special deflector designed to discharge most of water a considerable distance away from wall.	• When pressure at head is adequate, head will protect a wider area of coverage than standard sidewall head.	17/32" orifice 7.2 to 8.4	NFPA UL	• Spacing restrictions depend on width of room. • Considerable pressure at the head location. • U.L. listing.
Open Sprinklers	• No fusible element. • Visible 1/2" or 17/32" orifice. • Pendent or upright • SSP or SSU	• Deluge systems.	1/2" 5.2 to 5.8 17/32" orifice 8.0 (average)	NFPA UL FM	• Deluge system restrictions per NFPA and FM.
Corrosion Resistant Sprinklers	• Coated sprinkler heads with wax, lead, or special plating.	• Heads installed in areas subject to corrosion, or where heads are exposed to atmosphere.	All	NFPA FM UL	• Coating must be applied by sprinkler head manufacturer—not in field.

General Use Sprinklers (continued)

Sprinkler Type	Characteristics	Usage	"K" Factor	Reference	Restrictions
Dry Pendent Sprinklers	• Pendent head as integral part of drop nipple. • Cap at top of drop nipple.	• Heads installed in pendent position on dry pipe system.	5.6 average 1/2″ orifice	NFPA FM UL	• Dry pendent head and drop nipple manufactured as a unit in standard lengths, or can be manufactured to specified dimensions.
Dry Upright Sprinklers	Upright sprinkler on dry pendent sprinkler nipple.	• Used to install sprinklers in upright position on wet pipe system. • Pipe in heated area heads nippled up into unheated area.	5.6 average 1/2″ orifice	NFPA FM UL	• Seal of dry shaft in heated area pipe fitting.
Ornamental Sprinklers	• Sprinkler heads painted or plated. • Chrome and special plating. • Most painted colors.	• Heads in areas where appearance is important.	All	NFPA FM UL	• Paint and plating must be applied by manufacturer; heads must never be painted in the field.
Flush Sprinklers	• Sprinklers where all or part of body including shank thread-mounted above ceiling line.	• Heads in areas where appearance is important.	All	NFPA FM UL	Comply with NFPA and FM relative to deflector distance below ceiling.
Small Orifice Heads	• Pendent with small deflector with pintle extending below deflector. • Upright with large deflector with pintle extending above deflector.	• Special areas, usually light hazard, where less water discharge, density, and coverage are required.	1/4″ orifice 1.3 to 1.5 3/8″ orifice 2.6 to 2.9 7/16″ orifice 4.0 to 4.4	NFPA FM UL	• Insurance carrier requirements • NFPA restrictions. • FM restrictions.

Special Use Sprinklers

Sprinkler Type	Characteristics	Usage	"K" Factor	Reference	Restrictions
Large Drop Sprinkler	• Discharge consists of large drops. At specific pressure, they discharge 100% more than standard. • 1/2" orifice head and 40% more than 17/32" head. • Minimum operating pressure: 25 psi. 80 sq. ft. minimum • Head spacing, 130 sq. ft. maximum.	• High challenge fires with strong head plume updraft. • Primarily storage protection and rack storage. • Further testing will expand future usage.	11 to 11.5	NFPA FM UL	Use at this writing is rather restrictive, but as tests continue, usage will expand as it is further tested and approved. (Many miscellaneous restrictions on design and usage)
Early Suppression Fast Response Sprinklers (ESFR)	• Activated by heat five times faster than regular heads at 165°F.	• Primarily storage and rack storage. Also residential. • Testing and combining with large drop sprinklers will greatly expand usage.	5.2 to 5.8	NFPA FM UL	• Limited to protection where tests have resulted in approvals. • Use on wet systems. • Operating p.s.i. under test conditions 50 p.s.i. • (Many restrictions on design and usage)

Pipe Table

The selection of the proper pipe for suppression systems to carry water to the discharge devices deserves careful evaluation, not only of the environmental conditions where the pipe will be installed, but also the nature of combustion the hazard is capable of producing. In addition, careful attention must be given to the governing building codes, fire codes, UL (Underwriter's Laboratory) listings, and FM (Factory Mutual) approvals relative to the use of the selected pipe for a particular system and installation

Plastic pipe for sprinkler systems has many design and applications restrictions. When the use of plastic pipe is considered, UL listings, FM approvals, and NFPA requirements must be evaluated with great care. Building code requirements and restrictions must also be determined.

The following table describes in a brief, generic way, some of the various pipe and fitting types, and is presented to provide the reader with an overview of the pipe and fittings available. With plastic pipe, the restrictions, increased usage parameters, UL listings and FM approvals are changing rapidly, and manufacturers' claims must be coordinated with approving agency requirements.

Pipe Table

Pipe Type	Joining Method		Advantages	"C" Factor	Reference
Steel Pipe Schedule 30 Schedule 40	• Cast iron • Malleable iron	screwed flanged welded cut-grooved couplings	• Universal usage, all sizes, all systems • Black or galvanized	120 wet 100 dry 120 galv.	NFPA ASTM FM UL
Steel Pipe Schedule 10 (Thin wall)	• Cast iron • Malleable iron	flanged welded roll-grooved couplings	• Greater internal pipe diameter • Superior hydraulics • Smaller pipe sizing	120 wet 100 dry 120 galv.	NFPA ASTM FM UL
Threaded Lightwall	• Screwed		• Approximately Schedule 10 internal diameter except at threads. • Schedule 10 hydraulic advantage with threads.	120 wet 100 dry 120 galv.	NFPA ASTM FM UL Mfg.
Copper Pipe	• Fluxed and soldered 95-5 tin-antimony solder ASTM B-32		• Ease of installation • Lighter pipe • Labor economy • Flexibility (Note: restrictions on usage)	150	NFPA ASTM FM UL Mfg.
Plastic Pipe Post chlorinated Polyvinyl Chloride (CPVC) Pipe and Fittings	• Assembled with solvent cement		• Light weight • Ease of installation • Economy in retrofit • On-site fabrication • Superior hydraulics • Labor economy (Note: restrictions on usage)	150	NFPA ASTM FM UL Mfg.
Plastic Pipe Polybutylene Pipe and Fittings	• Jointed by heat fusion—with Mfg.'s heat fusion tool. No solvents, flux, pipe dope.		• Light weight • Ease of installation • Economy in retrofit • On-site fabrication • Superior hydraulics • Labor economy (Note: restrictions on usage)	150	NFPA ASTM FM UL Mfg.

INDEX